HRW
ALGEBRA ONE
INTERACTIONS
COURSE 2
RETEACHING MASTERS

$$f(\ell)=\frac{6600}{\ell}$$

$$a(t)=5.5t^2$$

HOLT, RINEHART AND WINSTON
Harcourt Brace & Company

Austin • New York • Orlando • Atlanta • San Francisco • Boston • Dallas • Toronto • London

To the Teacher

HRW Algebra One Interactions Course 2 Reteaching Masters contain two pages of alternative instruction strategies for the main skills in each lesson in *HRW Algebra One Interactions Course 2. Reteaching Masters* review the key vocabulary terms and use worked-out examples to provide instruction for the two or three most important skills in each lesson. A set of practice exercises in which students practice the precise skill taught follows the instruction.

Developmental assistance by B&B Communications West, Inc.

HRW is a registered trademark licensed to Holt, Rinehart and Winston, Inc.

Printed in the United States of America

ISBN 0-03-051292-1

1 2 3 4 5 6 7 066 00 99 98 97

TABLE OF CONTENTS

Reteaching
1.1 Addition and Subtraction Equations

◆ **Skill A** Simplifying expressions involving inverse operations

Recall Addition and subtraction undo each other.

◆ **Example 1**
Simplify the expression $12 - 8 + 8$.

◆ **Solution**
$12 - 8 + 8 = 12 + 0$ Additive Inverse
$12 + 0 = 12$ Simplify.

◆ **Example 2**
Write and simplify an expression to show that Mark deposited $20 to his existing bank balance and then withdrew $20.

◆ **Solution**
Let b = existing bank balance.
$b + 20 - 20 = b$

Simplify each equation.

1. $8 + 2 - 2$ _____

2. $19 - 7 + 7$ _____

3. $15 + 36 - 15$ _____

4. $-19 + 63 + 19$ _____

5. $75 + 27 - 27$ _____

6. $c + 9 - 9$ _____

7. $p - 14 + 14$ _____

8. $27 + m - 27$ _____

9. $x - 45 + 45$ _____

10. $t - 85 + 85$ _____

Write and simplify an expression to show each situation.

11. Maureen started her evening at the carnival with some money. She

won $22 playing games and then spent $22 on food and rides. _____

12. Dorothy added 5 points to her existing homework grade by doing extra

credit and then lost 5 points by forgetting to do her next assignment. _____

13. Anthony deducted $25 from his checking account for a bounced

check and then he deposited $25 in his account. _____

◆ **Skill B**   Writing addition and subtraction expressions

    Recall   The phrases *more than*, *exceeds*, and *is larger than* indicate addition.
                The phrases *less*, *is smaller than*, and *minus* indicate subtraction.

        ◆ **Example**
             Write an algebraic expression for the following verbal expression:
                       7 more than the number n

        ◆ **Solution**
            $n + 7$ or $7 + n$

Write an algebraic expression for each verbal expression.

14. 15 larger than the number c _____

15. the number t less 5 _____

16. 17 minus the number p _____

17. the number j plus 13 _____

◆ **Skill C**   Solving equations involving addition or subtraction

    Recall   Whatever amount you add to or subtract from one side of an equation, you must
                add to or subtract from the other side.

        ◆ **Example**
            Solve $p - 8 = 2$.

        ◆ **Solution**
$$p - 8 = 2$$
$$p - 8 + 8 = 2 + 8 \qquad \text{Add 8 to each side to undo subtraction.}$$
$$p = 10 \qquad \text{Simplify.}$$

Solve each equation.

18. $m - 11 = 23$ _____

19. $y + 0.6 = -1.3$ _____

20. $z + 7 = 20$ _____

21. $p - 14 = 11$ _____

22. $n - 1.9 = 3.1$ _____

23. $c + 24 = 37$ _____

24. $-79 + x = -64$ _____

25. $53 = r - 72$ _____

26. $2.5 + t = -3.52$ _____

27. $n - 33 = -21$ _____

28. $v - 31.7 = 42.1$ _____

29. $-52 + y = -2$ _____

30. $-21 + m = -72$ _____

31. $2.1 + p = 1.5$ _____

32. $z + 29 = 15$ _____

33. $17 + c = 23$ _____

Reteaching
1.2 Multiplication and Division Equations

◆ **Skill A** Determining the reciprocal of a number

Recall The reciprocal of a nonzero fraction is formed by inverting the fraction. If the number is a whole number, place it over a denominator of 1, and then invert it.

◆ **Example 1**
Find the reciprocal of $-\frac{3}{8}$.

◆ **Solution**
$-\frac{8}{3}$ Interchange the numerator and denominator.

◆ **Example 2**
Find the reciprocal of 5.

◆ **Solution**
$\frac{5}{1}$ Write 5 as the numerator over a denominator of 1.

$\frac{1}{5}$ Interchange the numerator and denominator.

Find the reciprocal of each number.

1. $-\frac{3}{4}$ _____

2. 3 _____

3. $\frac{1}{8}$ _____

4. -7 _____

5. $\frac{8}{5}$ _____

6. $-\frac{3}{2}$ _____

◆ **Skill B** Solving equations involving multiplication or division

Recall Whatever amount you multiply or divide one side of an equation by, you must multiply or divide the other side by.
The product of a nonzero number and its reciprocal is 1.

◆ **Example**
Solve $\frac{3}{2}x = -6$.

◆ **Solution**
$\frac{3}{2}x = -6$

$\frac{2}{3}\left(\frac{3}{2}x\right) = -6\left(\frac{2}{3}\right)$ Multiply each side by $\frac{2}{3}$, the reciprocal of $\frac{3}{2}$.

$x = -\frac{12}{3}$ Simplify.

$x = -4$

Solve each equation.

7. $-4y = 48$ _____

8. $\frac{x}{3} = -1.5$ _____

9. $-\frac{4}{3}w = -2$ _____

◆**Skill C**   Solving one-step equations involving percents

Recall   A percent can be expressed as a decimal by moving the decimal point two places to the left and dropping the percent sign.
Sale price is a percentage of the regular price.

◆ **Example**
A VCR is advertised at 60% of the regular price. If it is on sale for $132, what is the regular price?

◆ **Solution**
Let r = the regular price.

$132 = 0.60r$ Change 60% to a decimal.

$\dfrac{132}{0.60} = \dfrac{0.60r}{0.60}$ Undo the multiplication by dividing each side by 0.60.

$\$220 = r$ Simplify.

The regular price is $220.

Write an equation and solve.

10.   Gail bought a ring for $240 at a 20% off sale. What was the original price?   _____

11.   Chris found a surfboard on sale for $180. If this price represented 90% of the original price, what was the original price?   _____

◆**Skill D**   Solving equations involving proportions

Recall   The cross products in a proportion are equal.

◆ **Example**
The ratio of boys to girls in an algebra class is 3 to 5. If the number of boys is 9, how many girls are in the class?

◆ **Solution**
Write an equation of equal ratios.
Let g = number of girls in the class.

Then, $\dfrac{3}{5} = \dfrac{9}{g}$

$3g = 45$ The cross products $3 \cdot g$ and $5 \cdot 9$ are equal.

$\dfrac{3g}{3} = \dfrac{45}{3}$ Divide each side by 3 to undo the multiplication.

$g = 15$ Simplify.

There are 15 girls in the class.

Solve by using proportions.

12.   The ratio of nonfiction books to fiction books in the school library is 7 to 4. If the number of fiction books is 480, how many nonfiction books are there?   _____

NAME _____ CLASS _____ DATE _____

 **Reteaching**
1.3 Two-Step Equations

◆ **Skill A**   Solving two-step equations

Recall   When two operations are involved in an equation, undo addition or subtraction first, and then undo multiplication or division.

◆ **Example 1**
Solve $4x - 9 = 35$.

◆ **Solution**

$4x - 9 = 35$	The operations involved are multiplication and subtraction.
$4x - 9 + 9 = 35 + 9$	Undo the subtraction first.
$4x = 44$	Simplify.
$\frac{4x}{4} = \frac{44}{4}$	Undo the multiplication.
$x = 11$	Simplify.

Identify the operations involved in each equation.

1. $5y - 8 = 42$ _____

2. $2t + 7 = -13$ _____

3. $\frac{c}{11} + 5 = 13$ _____

4. $\frac{m}{4} - 12 = 15$ _____

5. $11 + 3p = 15$ _____

6. $-4 + \frac{n}{5} = 11$ _____

7. $4c - 20 = 80$ _____

8. $\frac{d}{2} + 14 = 32$ _____

9. $12t + 9 = 73$ _____

10. $\frac{r}{0.5} - 1.2 = 3.8$ _____

Solve each equation.

11. $2p - 11 = 9$ _____

12. $\frac{m}{4} + 5 = 11$ _____

13. $6r + 5 = 47$ _____

14. $14c - 12 = 72$ _____

15. $\frac{y}{7} - (-0.8) = 1.6$ _____

16. $-\frac{2}{3}x - 5 = 23$ _____

17. $\frac{1}{5}t - 11 = 22$ _____

18. $\frac{3}{5}c + 8 = 26$ _____

19. $\frac{7}{8}r - 2 = 12$ _____

20. $\frac{y}{6} + 5 = -11$ _____

21. $\frac{3}{5}p - 2 = \frac{2}{5}$ _____

22. $-\frac{2}{3}m + 7 = -11$ _____

23. $\frac{r}{2} - \left(-\frac{3}{2}\right) = 2$ _____

24. $6y - 25 = 5$ _____

25. $\frac{5}{8}t + \frac{3}{2} = 4$ _____

26. $-\frac{7}{3}c - 2 = \frac{1}{3}$ _____

HRW Algebra One Interactions Course 2   **Reteaching 1.3**   **5**

◆ **Skill B** Using geometric formulas to find values

Recall Substitute given values for the variables in the formula before solving.

◆ **Example 1**

The perimeter of a rectangle is found using the formula $P = 2l + 2w$, where P is the perimeter, l is the length, and w is the width of the rectangle. Find the length of a rectangle that has a width of 12 cm and a perimeter of 60 cm.

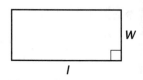

◆ **Solution**

$$P = 2l + 2w$$
$$60 = 2l + 2 \cdot 12 \quad \text{Substitute given values.}$$
$$60 = 2l + 24 \quad \text{Simplify.}$$
$$60 - 24 = 2l + 24 - 24 \quad \text{Undo addition first.}$$
$$36 = 2l$$
$$\frac{36}{2} = \frac{2l}{2} \quad \text{Undo multiplication.}$$
$$18 = l$$

The length is 18 cm.

◆ **Example 2**

The perimeter of an equilateral triangle is given by the formula $P = 3s$, where P is the perimeter and s is the length of a side. If the perimeter of an equilateral triangle is reduced by 4 inches, it becomes 47 inches. Find the length of a side of the triangle.

◆ **Solution**

$$P = 3s$$
$$47 = 3s - 4 \quad \text{Substitute the given values.}$$
$$47 + 4 = 3s - 4 + 4 \quad \text{Undo subtraction first.}$$
$$51 = 3s \quad \text{Simplify.}$$
$$\frac{51}{3} = \frac{3s}{3} \quad \text{Undo multiplication.}$$
$$17 = s$$

The length of a side is 17 inches.

Use the formula $P = 2l + 2w$ with the given values to find the missing value.

27. $P = 76, l = 17$ _____

28. $P = 54, w = 7$ _____

29. $P = 46, w = 10$ _____

30. $P = 100, l = 30$ _____

Use the formula $P = 3s$ to solve each of the following:

31. If the perimeter of an equilateral triangle is reduced by 13 feet, the new perimeter becomes 47 feet. Find the length of a side of the original triangle. _____

32. If the perimeter of an equilateral triangle is increased by 12 meters, its new perimeter becomes 48 meters. Find the length of a side of the original triangle. _____

Reteaching
1.4 Multistep Equations

◆ **Skill A** Combining like terms

Recall Like terms have the same variables, including exponents. Like terms can be combined by using the Distributive Property.

◆ **Example**
Combine $4x^2 - 2x^2 + 7x - 11x - 12$.

◆ **Solution**
$(4 - 2)x^2 + (7 - 11)x - 12$ Rewrite each group of like terms by using the
$2x^2 + (-4x) - 12$ Distributive Property.

Combine each of the like terms.

1. $2p + 10p - 4p$ _____

2. $11m - 7m + 6m$ _____

3. $3x^2 + 12x^2$ _____

4. $4y - 7y + 2y$ _____

5. $15y^2 - 3y^2 + 4y^2$ _____

6. $9n + 12n - 13n$ _____

7. $3m^2 - m^2 + 3m + 5m$ _____

8. $7t^2 - 9t^2 + 3t + 5t - 1$ _____

9. $12c^2 + 4c + 7c^2 - 6c + 2$ _____

10. $15r^2 - 21r^2 - 7r - 12 + 3r$ _____

◆ **Skill B** Solving multistep equations by grouping like terms

Recall The Addition and Subtraction Properties of Equality allow you to add or subtract the same, or equal, quantities from each side of an equation.

◆ **Example**
Solve $5x - 8 = 12 - 3x$.

◆ **Solution**
$5x - 8 = 12 - 3x$
$5x + 3x - 8 = 12 - 3x + 3x$ Add $3x$ to each side to group like terms.
$8x - 8 = 12$ Combine like terms.
$8x - 8 + 8 = 12 + 8$ Add 8 to each side.
$8x = 20$ Simplify.
$\dfrac{8x}{8} = \dfrac{20}{8}$ Divide each side by 8.
$x = 2.5$ Simplify.

Solve each equation.

11. $4y - 12 + 3y + 18 = 27$ _____

12. $-3x = 143 + 8x$ _____

13. $0.6x + 1 = -0.2x + 3.2$ _____

14. $63w + 11w = 52w + 121$ _____

◆ **Skill C**   Removing parentheses preceded by a negative sign

     Recall   Multiplication by a positive number does not change a factor's value.

           Multiplication by a negative number produces the opposite of the original factor when the parentheses are removed.

       ◆ **Example 1**
           Simplify $-(x - 2)$.

       ◆ **Solution**

$-(x - 2) = -1(x - 2)$	Replace the negative sign with -1.
$\quad\quad\quad = -1x - 1(-2)$	Use the Distributive Property.
$\quad\quad\quad = -x + 2$	Simplify.

       ◆ **Example 2**
           Simplify $3(x - 2) - (2x - 4)$.

       ◆ **Solution**

$3(x - 2) - (2x - 4)$	Use the Distributive Property.
$3x - 6 - 2x + 4$	The first sign does not change. The second sign does change.
$3x - 2x - 6 + 4$	Group like terms.
$x - 2$	Combine like terms.

Simplify.

15. $-(y + 4)$ _____   **16.** $-(c - 5)$ _____

17. $-(-t + 2)$ _____   **18.** $-(-m - 4)$ _____

19. $2(x - 1) - (x + 5)$ _____   **20.** $5(x + 4) - (x - 3)$ _____

21. $-3(x - 4) + 2(2x - 3)$ _____   **22.** $-(2x + 3) + 4(x + 2)$ _____

23. $-(x - 2) - 2(x + 2)$ _____   **24.** $-2(-3m - 2) - 4(m + 3)$ _____

Solve each equation.

25. $5(x - 3) - (x - 7) = 32$ _____   **26.** $7(y + 3) - (y - 11) = 50$ _____

27. $-4(p + 12) + (p - 9) = 15$ _____   **28.** $6(r - 8) - (2r - 3) = 55$ _____

29. $2(8h + 15) - (h + 6) = 99$ _____   **30.** $12(2q - 3) - 3(q + 15) = 24$ _____

31. $-3(2x - 5) - 2(-x - 5) = -3$ _____   **32.** $-4(2s + 3) - 3(-2s - 4) = -3$ _____

33. $4t - t(-3 - 5) - 2(t - 3) = 6$ _____   **34.** $-5c - 2(3c - 5) + 2 = -10$ _____

35. $-5(-g - 4) - 2(5 - 3g) = -1$ _____   **36.** $7(2v - 3) - 3(-2v - 10) = 1$ _____

Reteaching
1.5 Solving Inequalities

◆ **Skill A**   Interpreting expressions involving inequality symbols

Recall   The symbol for *less than* points to the left: <.
The symbol for *greater than* points to the right: >.
At least means "greater than or equal to." *At most* means "less than or equal to."

◆ **Example**
Write an inequality to represent the following: The cost of a ticket to the school dance is $3. The expenses are $760. How many tickets must be sold so that there will be a profit of at least $500?

◆ **Solution**
Profit = amount taken in minus expenses
Let t = number of tickets; then $3t$ is the total value of tickets sold.
$3t - 760 \geq 500$

Write an inequality to represent each of the following:

1. The product of 2 and a number is greater than 8. _____

2. The quotient of 5 and a number is less than or equal to -11. _____

3. The product of a number and 16 is less than five. _____

4. The quotient of a number and 13 is greater than or equal to 52. _____

5. The product of 2 times a number and 4 is less than or equal to a

   number plus 8. _____

6. To receive an A, Jeremy must earn at least 670 points. So far, he has 520 points, with two more 100-point quizzes to take. What must he receive as an average score on each quiz to earn an A?

   _____

◆ **Skill B** Graphing on the number line

Recall A value that is included is represented by a filled-in circle. A value that is not included is represented by an open circle. *Greater than* is indicated by shading the points on the number line to the right. *Less than* is indicated by shading the points to the left.

◆ **Example**
Graph the solution on a number line.
a. $x \geq 3$ **b.** $y < -2$

◆ **Solution**
a.
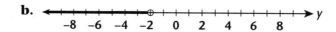

b.

Graph each inequality on the number line provided.

7. $t \geq 4$

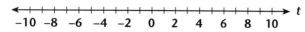

8. $m < 5$

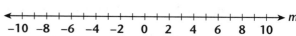

◆ **Skill C** Solving inequalities

Recall Adding or subtracting the same quantity from both sides of an inequality does not change the direction of the inequality symbol.

◆ **Example 1**
Solve $4x - 3 \leq 13$.

◆ **Solution**
$4x - 3 \leq 13$
$4x \leq 16$
$x \leq 4$ Simplify.

Recall Multiplying or dividing by a negative quantity on both sides of an inequality does change the direction of the inequality symbol.

◆ **Example 2**
Solve $-3y + 3 \geq 15$.

◆ **Solution**
$-3y + 3 \geq 15$
$-3y \geq 12$
$y \leq -4$ Dividing both sides by -3 reverses the direction of the inequality symbol.

Solve each inequality.

9. $5x - 7 \geq 28$ _____

10. $-4t - 13 < 35$ _____

11. $-2p - 8 > 14$ _____

12. $6y + 8 \leq 20$ _____

Reteaching
1.6 Absolute-Value Equations and Inequalities

◆ **Skill A** Solving absolute-value equations

Recall The distance that a number is from zero on a number line does not involve direction. Therefore, a given distance from zero will always have two values.

◆ **Example 1**
Solve $|c| = 8$.

◆ **Solution**
Since $|8| = 8$ and $|-8| = 8$, c can be either 8 or -8.

◆ **Example 2**
Solve $|y - 3| = 11$.

◆ **Solution**
Since $|11| = 11$ and $|-11| = 11$, $y - 3$ can be either 11 or -11.

Case 1   Case 2
$y - 3 = 11$ $y - 3 = -11$
$y = 14$ $y = -8$

For each absolute-value equation, name the two values that the expression can equal.

1. $|t| = 12$ _____
2. $|m| = 5$ _____
3. $|s| = 24$ _____
4. $|p| = 5.3$ _____
5. $|y + 4| = 11$ _____
6. $|i - 3| = 15$ _____
7. $|z| = 14$ _____
8. $|t| = 36$ _____

Solve each absolute-value equation.

9. $|x - 1.1| = 1.7$ _____
10. $|c + 7| = 45$ _____
11. $|y - 6| = 22$ _____
12. $|x + 5| = 18$ _____
13. $|r - 1.1| = 3.8$ _____
14. $|p + 9| = 29$ _____
15. $|q + 8| = 70$ _____
16. $|d - 1| = 62$ _____
17. $|a - 5| = 2.3$ _____
18. $|x + 22| = 7.5$ _____
19. $|t + 25| = 5$ _____
20. $|p - 4.3| = 5.7$ _____
21. $|r - 10.1| = -2.3$ _____
22. $|q + 72| = 3$ _____

◆ **Skill B** Writing and solving absolute-value inequalities

Recall If the range cannot exceed a given value, then the range must be less than or equal to the value.

◆ **Example**
Prizes are given to students who guess within a given range the number of jelly beans in a jar. The number of jelly beans in the jar is 164, and prizes are given to those who guess no more than 6 from the actual amount. Write an equation to represent the range for the winning guesses. Then solve the equation.

◆ **Solution**
Since the guessed amount can be either 6 above or 6 below the actual amount, an absolute-value inequality can be used to represent the problem. The difference between the guess and the actual number must be less than or equal to 6.

Let x = the guess.
$$|x - 164| \leq 6$$

$$x - 164 \leq 6 \qquad\qquad x - 164 \geq -6$$
$$x \leq 170 \qquad\qquad x \geq 158$$

The number guesses must be between 158 and 170.

Write an equation to represent each situation. Then solve the equation.

23. On a television game show, a contestant will win a new car if he or she can guess its actual value within $100. If the actual value is $13,200, what is the range within which the contestant can guess?

_____

_____

24. The cooking time for a given turkey is 310 minutes, although any time within 12 minutes of the given time is considered acceptable. What range of cooking times is acceptable?

_____

_____

25. Jen should consume 1800 calories per day to maintain her weight. She has found that if she consumes any number of calories within 50 of her optimal number, she will maintain her present weight. What is the acceptable calorie range?

_____

_____

Reteaching
2.1 Exploring Slope

◆ **Skill A**   Finding the slopes of lines from their graphs

Recall   Slope is the ratio of vertical change to horizontal change between two points on a line. A line that slants upward from left to right has positive slope. A line that slants downward from left to right has a negative slope. A horizontal line has a slope of zero. The slope of a vertical line is undefined.

◆ **Example 1**
Determine whether the slope of the line is positive or negative.

◆ **Solution**
Because the line slants downward from left to right, the slope is negative.

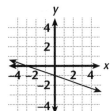

◆ **Example 2**
Determine the slope of the line.

◆ **Solution**
Choose two points on the line whose coordinates are integers. Draw a line from one point to the other. Find the horizontal distance and the vertical distance by counting the boxes. Form the ratio of vertical distance to horizontal distance. Determine whether the slope is positive or negative.

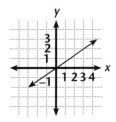

$$\text{slope} = \frac{\text{vertical distance}}{\text{horizontal distance}}$$

$$= \frac{2}{3}$$

The line slants upward from left to right, so the slope is $+\frac{2}{3}$.

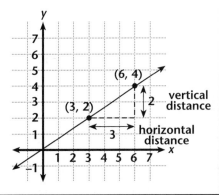

Determine whether the slope of each line is positive, negative, zero, or undefined.

1.

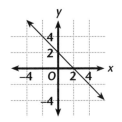

2.

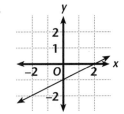

3.

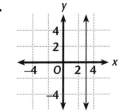

_____   _____   _____

Find the slope of each line.

4.

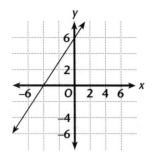

_____

5.

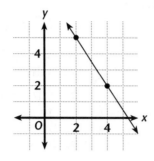

_____

6.

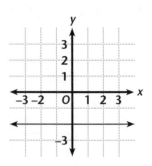

_____

◆ **Skill B** Determining the slope of a line through two given points

Recall slope $= \dfrac{\text{difference in } y\text{-coordinates}}{\text{difference in } x\text{-coordinates}}$

Be sure to carry out the subtraction in the same order.

◆ **Example**
Determine the slope of the line through the points $(-2, 3)$ and $(-4, -1)$.

◆ **Solution**

slope $= \dfrac{\text{difference in } y\text{-coordinates}}{\text{difference in } x\text{-coordinates}}$

$(-2, 3)$ $(-4, -1)$

difference in x-coordinates   difference in y-coordinates

$(-4) - (-2)$ $(-1) - (3)$

slope $= \dfrac{\text{difference in } y\text{-coordinates}}{\text{difference in } x\text{-coordinates}}$

$= \dfrac{-1 - 3}{-4 - (-2)}$

$= \dfrac{-4}{-2}$

$= 2$

Determine the slope of the line through the given points.

7. $(5, 3)$ and $(2, 7)$ _____

8. $(-1, 0)$ and $(-4, 3)$ _____

9. $(-2, 6)$ and $(4, 7)$ _____

10. $(0, 7)$ and $(-4, 9)$ _____

11. $(4, 5)$ and $(-7, -4)$ _____

12. $(8, -6)$ and $(-5, 3)$ _____

Reteaching
2.2 Linear Functions and Graphs

◆ **Skill A** Identifying the slope and y-intercept from equations of lines

Recall The slope-intercept form for the equation of a line is $y = mx + b$, where m is the slope and b is the y-intercept.

The standard form for the equation of a line is $Ax + By = C$, where A, B, and C are integers, A and B are not both zero, and A is not negative.

◆ **Example**
Identify the slope and y-intercept of each line.

    a. $y = -3x - 5$ **b.** $3x + 4y = 8$

◆ **Solution**
    a. Rewrite the equation so that it is in the form $y = mx + b$.

$$y = -3x - 5$$
$$y = -3x + (-5)$$

The slope is -3 and the y-intercept is -5.

    b. The equation is in standard form. Rewrite it so that it is in slope-intercept form.

$$3x + 4y = 8$$

$$3x + 4y - 3x = 8 - 3x \qquad \text{Subtract } 3x \text{ from each side.}$$
$$\tfrac{4}{4}y = -\tfrac{3}{4}x + \tfrac{8}{4} \qquad \text{Divide each side by 4.}$$
$$y = -\tfrac{3}{4}x + 2 \qquad \text{Simplify.}$$

The slope is $-\tfrac{3}{4}$ and the y-intercept is 2.

Identify the slope and y-intercept of each line.

1. $y = -\tfrac{1}{3}x + 6$ _____

2. $y = x - 1$ _____

3. $3x + 5y = 15$ _____

4. $y = -2x - 5$ _____

5. $-2x + 4y = -8$ _____

6. $5x - 2y + 4 = 0$ _____

7. $y = \tfrac{2}{3}x + 3$ _____

8. $-x - 3y = 9$ _____

9. $y = 3x$ _____

10. $y = -6$ _____

◆ **Skill B** Graphing the equation of a line by using the slope and y-intercept

Recall slope $= m = \dfrac{\text{vertical distance}}{\text{horizontal distance}}$

◆ **Example**
Draw the graph of $y = \frac{3}{5}x - 4$.

◆ **Solution**
Use the y-intercept as the starting point.
Use the slope to find a second point on the line.
Connect the two points with a straight line.

The y-intercept is $(0, -4)$ and the slope is $\frac{3}{5}$.

Start at $(0, -4)$. Move up 3 units and to the
right 5 units to find another point on the line.
Draw a line through the y-intercept and the
new point, $(5, -1)$.

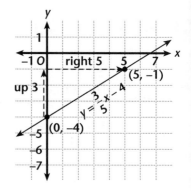

Draw the graph of each line on graph paper.

11. $y = \frac{3}{2}x - 3$

12. $y = \frac{1}{2}x + 2$

13. $y = -4x + 1$

14. $y = 2x - 2$

15. $2x - 3y = 6$

16. $-5x + 4y = 12$

◆ **Skill C** Writing equations given two points on the line

Recall $m = \dfrac{\text{change in } y}{\text{change in } x}$

◆ **Example**
Write an equation of the line through $(3, 7)$ and $(8, 17)$.

◆ **Solution**
First find the slope to find the value of m in $y = mx + b$.

$$m = \frac{\text{change in } y}{\text{change in } x} = \frac{17 - 7}{8 - 3} = \frac{10}{5} = 2$$

Use 2 for m and one of the points to find the value of b.
$y = 2x + b$
$7 = 2(3) + b$ ⠀⠀⠀⠀⠀Substitute 7 for y and 3 for x.
$7 = 6 + b$
$1 = b$

The equation is $y = 2x + 1$.

**Write an equation in the slope-intercept form of the line
through each of the two given points.**

17. $(5, -6)$ and $(3, -9)$ _____

18. $(-4, 0)$ and $(0, 12)$ _____

19. $(0, 120)$ and $(2, 160)$ _____

20. $(10, 0)$ and $(-3, -26)$ _____

Reteaching
2.3 Graphing Systems of Equations

◆ **Skill A** Determining whether given points are on a given line

Recall If a point is on a line, when the coordinates of the point are substituted into the
equation, the left and right sides of the equation will be equal.

◆ **Example**
Determine whether the point $(-2, 5)$ is on the line $4x + 3y = 7$.

◆ **Solution**
Substitute -2 for x and 5 for y.
$$4x + 3y \overset{?}{=} 7$$
$$4(-2) + 3(5) \overset{?}{=} 7$$
$$-8 + 15 \overset{?}{=} 7$$
$$7 = 7 \checkmark$$
Since the left and right sides are equal, the point lies on the line.

Determine whether each given point lies on the given line.

1. $(2, 5)$, $2x - 3y = -11$ _____

2. $(4, -3)$, $5x + 2y = 7$ _____

3. $(4, 2)$, $-2x - 3y = -14$ _____

4. $(-6, 4)$, $-2x + y = 16$ _____

◆ **Skill B** Finding the intersection of the graphs of two linear equations

Recall If a point (a, b) is the intersection of the graphs of two linear equations, then the
solution to the system is $x = a$ and $y = b$.

◆ **Example**
Find the point of intersection of the two linear graphs. Show that the
coordinates of that point are the solution to the system.

◆ **Solution**
The graphs of $y = x + 2$ and $2x + y = 5$
are shown at the right and *appear* to
intersect at $(1, 3)$.

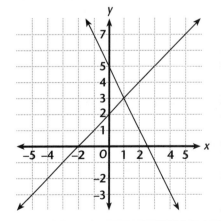

$$y = x + 2 \qquad\qquad 2x + y = 5$$
$$3 \overset{?}{=} 1 + 2 \qquad\qquad 2(1) + 3 \overset{?}{=} 5$$
$$3 = 3 \quad \checkmark \qquad\qquad 2 + 3 \overset{?}{=} 5$$
$$\qquad\qquad\qquad\qquad 5 = 5 \quad \checkmark$$
The point $(1, 3)$ is on both graphs.
The solution to the system is $x = 1$
and $y = 3$.

Find the point of intersection and show that the coordinates of that point are the solution to each equation.

5.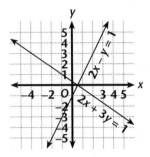

6.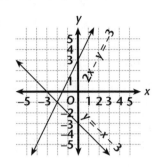

◆ **Skill C**   Solving a system of equations by graphing

Recall   The solution to a system of equations is the point where the graphs intersect.

◆ **Example**
Solve the system of equations by graphing.
$$\begin{cases} 2x - 3y = 6 \\ x + 2y = 3 \end{cases}$$

◆ **Solution**
Use the method of your choice to graph both equations.

$2x - 3y = 6$ $\qquad\qquad\qquad\qquad$ $x + 2y = 3$

Write in slope-intercept form.

Let $x = 0$.
$$0 - 3y = 6$$
$$y = -2$$

Let $y = 0$.
$$2x - 0 = 6$$
$$x = 3$$

$(0, -2)$ and $(3, 0)$ are on the graph of $2x - 3y = 6$.

$2y = -x + 3$
$$y = -\frac{1}{2}x + \frac{3}{2}$$

The y-intercept is $\frac{3}{2}$; the slope is $-\frac{1}{2}$.

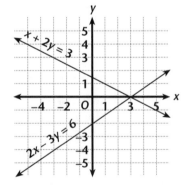

The graphs appear to intersect at $(3, 0)$.

$2x - 3y = 6$	$x + 2y = 3$
$2(3) - 3(0) = 6$	$3 + 2(0) = 3$
$6 - 0 = 6$	$3 + 0 = 3$
$6 = 6$	$3 = 3$

The solution is $x = 3$ and $y = 0$.

Solve each system of equations by graphing.

7. $\begin{cases} x - y = 2 \\ 2x + 3y = 9 \end{cases}$ _____

8. $\begin{cases} 5x + 3y = 12 \\ x + 2y = 8 \end{cases}$ _____

9. $\begin{cases} y = x + 3 \\ x + y = 9 \end{cases}$ _____

10. $\begin{cases} 2x + y = 0 \\ 3x - y = 5 \end{cases}$ _____

 **Reteaching**
2.4 Exploring Substitution Methods

◆ Skill A Substituting an expression for a variable in equations

Recall Use the Distributive Property to simplify equations with parentheses.

◆ **Example**
Substitute the expression $3x - 2$ for y in the equation $2x + 3y = 5$.
Find the value of x.

◆ **Solution**
Replace the y-variable with the expression $3x - 2$.

$$2x + 3y = 5$$
$$2x + 3(3x - 2) = 5 \qquad \text{Substitute } 3x - 2 \text{ for } y.$$
$$2x + 9x - 6 = 5 \qquad \text{Use the Distributive Property.}$$
$$11x - 6 = 5 \qquad \text{Combine like terms.}$$
$$11x = 11$$
$$x = 1$$

Substitute the given expression for the same variable in the second equation. Solve for the remaining variable.

1. $y = 2x + 4$ _____
   $3x + 2y = 29$

2. $y = 5x + 2$ _____
   $4x + 3y = 25$

3. $x = -3y + 3$ _____
   $2x + 4y = 8$

4. $x = 3y - 5$ _____
   $3x + 3y = 9$

5. $y = 3x + 1$ _____
   $-2x - 3y = 19$

6. $x = -2y + 1$ _____
   $3x + 8y = 9$

7. $y = 5x - 1$ _____
   $x + y = 5$

8. $x = 4y + 3$ _____
   $3x + 6y = 27$

◆ **Skill B**   Solve a system of equations by substitution

Recall   The solution to a system of equations must satisfy all equations in the system. Isolating a variable means having the variable with a coefficient of 1 alone on one side of the equation.

◆ **Example**

Determine the solution to each system of equations by substitution.

a. $\begin{cases} y = x - 3 \\ 2x + 3y = 16 \end{cases}$
b. $\begin{cases} x + y = 6 \\ 3x - 4y = 4 \end{cases}$

◆ **Solution**

a.   Substitute $x - 3$ for y in the second equation.

$$2x + 3y = 16$$
$$2x + 3(x - 3) = 16 \qquad \text{Use the Distributive Property.}$$
$$5x - 9 = 16 \qquad \text{Combine like terms.}$$
$$5x = 25$$
$$x = 5$$

To solve for y, substitute 5 for x in one of the equations.
For $y = x - 3$:
$$y = 5 - 3$$
$$y = 2$$
The solution is (5, 2).

Check the solution in the other equation. Substitute 5 for x and 2 for y in $2x + 3y = 16$.

$$2(5) + 3(2) \stackrel{?}{=} 16$$
$$10 + 6 \stackrel{?}{=} 16$$
$$16 = 16$$

b.   Since neither equation has a variable isolated, choose one. The easier choice is the first equation because both variables have coefficients of 1.

$$x + y = 6$$
$$y = 6 - x$$

Substitute for y in the second equation.

$$3x - 4y = 4$$
$$3x - 4(6 - x) = 4$$
$$3x - 24 + 4x = 4$$
$$7x - 24 = 4$$
$$7x = 28$$
$$x = 4$$

Substitute 4 for x in the equation $x + y = 6$.
$$4 + y = 6$$
$$y = 2$$
The solution is $x = 4$ and $y = 2$.

Solve each system by substitution.

9. $\begin{cases} y = 2x \\ 3x + 2y = 21 \end{cases}$ _____

10. $\begin{cases} y = 6 - x \\ 9x - 5y = 40 \end{cases}$ _____

11. $\begin{cases} y - x = 2 \\ 2x + 3y = 21 \end{cases}$ _____

12. $\begin{cases} x - y = -3 \\ 2x + y = 12 \end{cases}$ _____

NAME _____ CLASS _____ DATE _____

 **Reteaching**
2.5 The Elimination Method

◆ Skill A Finding the opposite of a given term

Recall The sum of opposites is zero.

◆ **Example**
Find the opposite of each term.
a. $5x$ b. $-7y$

◆ **Solution**
a. $-5x$ is the opposite of $5x$ because $5x + (-5x) = 0$.
b. $7y$ is the opposite of $-7y$ because $-7y + 7y = 0$.

Find the opposite of each term.

1. $-2t$ _____ 2. $5p$ _____ 3. $8c$ _____ 4. $-3m$ _____

◆ Skill B Finding the lowest common multiple (LCM)

Recall The LCM of two numbers is the smallest number that both numbers divide evenly.

◆ **Example 1**
Find the LCM of 4 and 5.

◆ **Solution**
List the multiples of 4: 4, 8, 12, 16, 20, 24, 28, ...
List the multiples of 5: 5, 10, 15, 20, 25, 30, 35, ...
The smallest number common to both lists is 20.

◆ **Example 2**
Find the LCM of 6 and 8.

◆ **Solution**
Find the prime factorization of 6: $2 \cdot 3$
Find the primer factorization of 8: $2 \cdot 2 \cdot 2 = 2^3$
To find the LCM, use each prime factor the greatest number of times it occurs in either number.
$\quad$ LCM $= 2^3 \cdot 3$ $\qquad$ 3 is used as a factor once in 6.
$\qquad\qquad\qquad\qquad\qquad$ 2 is used as a factor 3 times in 8.
$\quad$ LCM $= 2^3 \cdot 3 = 24$

Find the LCM of each pair by listing multiples.

5. $4, 9$ _____ 6. $3, 5$ _____ 7. $6, 10$ _____ 8. $4, 10$ _____

Find the LCM of each pair by prime factorization.

9. $10, 15$ _____ 10. $8, 12$ _____ 11. $20, 25$ _____ 12. $16, 24$ _____

HRW Algebra One Interactions Course 2   **Reteaching 2.5** **21**

◆ **Skill C** Solving a system of equations by using elimination

Recall If the coefficients of one of the variables are opposites, add the equations to eliminate one variable. If necessary, find the LCM of the coefficients of one variable, and then use the Multiplication Property to change the coefficients to opposites.

◆ **Example**

Solve each system by using the elimination method.

a. $\begin{cases} a - 3b = 1 \\ 2a + 3b = 20 \end{cases}$
    b. $\begin{cases} 2x + 6y = 18 \\ 4x - 3y = 6 \end{cases}$

◆ **Solution**

a. The coefficients of b are opposites. Add the two equations to eliminate the b-term.

$$\begin{array}{r} a - 3b = 1 \\ + 2a + 3b = 20 \\ \hline 3a \quad\quad = 21 \\ a = 7 \end{array}$$

Substitute 7 for a in either equation and solve for b.

$$\begin{array}{r} a - 3b = 1 \\ -3b = -6 \\ b = 2 \end{array}$$

The solution is $a = 7$ and $b = 2$.

b. None of the coefficients are opposites. Find the LCM of the coefficients of one variable. Multiply the second equation by 2 to make the y-coefficient -6. Then add.

$$2x + 6y = 18$$
$$2(4x) - 2(3y) = 2(6)$$

$$\begin{array}{r} 2x + 6y = 18 \\ 8x - 6y = 12 \\ \hline 10x \quad\quad = 30 \\ x = 3 \end{array}$$

Substitute 3 for x in one of the original equations to solve for y.

$$\begin{array}{r} 2x + 6y = 18 \\ 2(3) + 6y = 18 \\ 6 + 6y = 18 \\ 6y = 12 \\ y = 2 \end{array}$$

The solution is $x = 3$ and $y = 2$.

Solve each system by using the elimination method.

13. $\begin{cases} 3x + y = 10 \\ 4x - y = 4 \end{cases}$ _____

14. $\begin{cases} 4x + 5y = 41 \\ 7x + 5y = 53 \end{cases}$ _____

15. $\begin{cases} 2a - b = 8 \\ a + 2b = 9 \end{cases}$ _____

16. $\begin{cases} 3x + 5y = 27 \\ 2x + 3y = 17 \end{cases}$ _____

NAME _____ CLASS _____ DATE _____

 **Reteaching**

2.6 Non-Unique Solutions

◆ **Skill A** Identifying consistent and inconsistent systems of equations

Recall Parallel lines have the same slope but different y-intercepts.
Nonparallel lines have different slopes. Their y-intercepts may be the same or different.
The equations of parallel lines form an inconsistent system. Their graphs have no points in common.
The equations of nonparallel lines form a consistent system. Their graphs have at least one point in common.

◆ **Example**
Determine whether the system is inconsistent or consistent.

a. $\begin{cases} 2x + y = 10 \\ -4x - 2y = -12 \end{cases}$ b. $\begin{cases} -3x + 4y = 7 \\ 6x - 4y = 2 \end{cases}$

◆ **Solution**

a. Write each equation in slope-intercept form. Then compare their slopes and y-intercepts.

$$2x + y = 10 \qquad\qquad -4x - 2y = -12$$
$$y = -2x + 10 \qquad\qquad -2y = 4x - 12$$
$$\qquad\qquad\qquad\qquad y = -2x + 6$$

Since the lines have the same slope but different y-intercepts, they are parallel and the system is inconsistent.

b. Write each equation in slope-intercept form. Then compare their slopes and y-intercepts.

$$-3x + 4y = 7 \qquad\qquad 6x - 4y = 2$$
$$4y = 3x + 7 \qquad\qquad -4y = -6x + 2$$
$$y = \frac{3}{4}x + \frac{7}{4} \qquad\qquad y = \frac{3}{2}x - \frac{1}{2}$$

Since the lines have different slopes, they are not parallel. The equations of nonparallel lines form a consistent system.

Determine whether each system is inconsistent or consistent.

1. $\begin{cases} 6x - 2y = -10 \\ -3x + y = 8 \end{cases}$ _____

2. $\begin{cases} 4y + 3x = 11 \\ 9y - 7x = 11 \end{cases}$ _____

3. $\begin{cases} x + 5y = -15 \\ -x - 5y = 25 \end{cases}$ _____

4. $\begin{cases} 3x - y = 8 \\ x + 2y = -7 \end{cases}$ _____

◆ **Skill B** Identifying dependent and independent systems of equations

Recall A consistent system can be independent or dependent.

An independent system of equations has one point in common. Their graphs are intersecting lines.
A dependent system of equations has infinitely many points in common. Their graphs are the same line.

◆ **Example 1**
Determine whether the system is independent, dependent, or inconsistent.
$$\begin{cases} 3x - y = 5 \\ 6x = 2y + 10 \end{cases}$$

◆ **Solution**
Compare the equations by writing each in slope-intercept form.

$$3x - y = 5 \qquad\qquad 6x = 2y + 10$$
$$y = 3x - 5 \qquad\qquad 2y = 6x - 10$$
$$\qquad\qquad\qquad\qquad y = 3x - 5$$

Since the equations are the same, the system is dependent.

◆ **Example 2**
Determine the number of solutions to the system
$$\begin{cases} x + 2y = -4 \\ x - y = -3 \end{cases}$$

◆ **Solution**
Compare the equations by writing each in slope-intercept form.

$$x + 2y = -4 \qquad\qquad x - y = -3$$
$$2y = -x - 4 \qquad\qquad y = x + 3$$
$$y = -\frac{1}{2}x - 2$$

Since the equations have different slopes, their graphs are not parallel and intersect at one point. There is one solution to the system.

Determine whether each system is independent, dependent, or inconsistent.

5. $\begin{cases} 5x - 2y = 10 \\ -10x + 4y = 12 \end{cases}$ _____

6. $\begin{cases} 4y - 6x = 12 \\ 3x - 2y = -6 \end{cases}$ _____

7. $\begin{cases} x - y = 7 \\ x + y = 4 \end{cases}$ _____

8. $\begin{cases} 9x = -6y + 12 \\ 3x + 2y - 4 = 0 \end{cases}$ _____

Determine the number of solutions to each system.

9. $\begin{cases} 2x + 3y = 0 \\ 5x - 3y = 21 \end{cases}$ _____

10. $\begin{cases} -3x = 6y - 9 \\ 5x + 10y = 10 \end{cases}$ _____

11. $\begin{cases} x + y = -9 \\ 3x + 27 = -3y \end{cases}$ _____

12. $\begin{cases} x - 3y = 1 \\ 2x + 3y = 20 \end{cases}$ _____

Reteaching
2.7 Graphing Linear Inequalities

◆ **Skill A** Determining the boundary lines of linear inequalities

Recall If the inequality is ≥ or ≤, use a solid line to show that the boundary line is included in the solution.
If the inequality is > or <, use a dashed line to show that the boundary line is not included in the solution.

◆ **Example**
Identify whether the boundary line for each inequality is solid or dashed.

    a. $y \leq -\frac{2}{3}x + 5$ **b.** $3x - 2y > 7$

◆ **Solution**
    a. Since the inequality is ≤, the boundary line is solid.
    b. Since the inequality is >, the boundary line is dashed.

Identify whether the boundary line for each inequality is solid or dashed.

1. $x - 3y < 4$ _____   **2.** $5x + y \geq -1$ _____

3. $y \leq 3$ _____   **4.** $y > \frac{1}{3}x - 2$ _____

◆ **Skill B** Graphing solutions of linear inequalities

Recall The x-intercept of a line is the point where $y = 0$.
The y-intercept of a line is the point where $x = 0$.
If the coordinates of a point make an inequality true, then the point is in the solution of the inequality.

◆ **Example**
Graph the inequality $3x - 4y > 12$ to find its solution.
First find the intercepts of the boundary line.

Let $y = 0$.   Let $x = 0$.
$3x - 4(0) = 12$ $3(0) - 4y = 12$
$3x = 12$ $-4y = 12$
$x = 4$ $y = -3$

Use the intercepts to graph the boundary line.
Since the inequality is >, the boundary line is dashed.
Choose a point and see if its coordinates satisfy the inequality.

For $(0, 0)$: $3(0) - 4(0) \overset{?}{>} 12$
$0 \not> 12$

Since $(0, 0)$ does not make the inequality true, shade the half-plane that does not contain $(0, 0)$.

Use graph paper to find the solution of each inequality.

5. $2x + 3y \leq 9$ **6.** $3x + 3y > 6$ **7.** $y \geq \frac{3}{2}x - 2$

8. $-3x + 2y < 8$ **9.** $-2x - 4y > 12$ **10.** $-4x - 3y \leq -6$

◆ **Skill C** Graphing solutions of systems of linear inequalities

Recall The solution of a system of inequalities consists of all points in the plane where the solutions of each inequality overlap.

◆ **Example**
Solve by graphing: $\begin{cases} 2x + y \geq 1 \\ -3x + y < 4 \end{cases}$

◆ **Solution**
Graph each boundary line.

$2x + y = 1$

x	y
0	1
2	-3

$-3x + y = 4$

x	y
0	4
-2	-2

The boundary line for $2x + y \geq 1$ is solid because the inequality is $\geq$. The boundary line for $-3x + y < 4$ is dashed because the inequality is $<$.
Decide which half-plane to shade for each inequality.
For (0, 0):

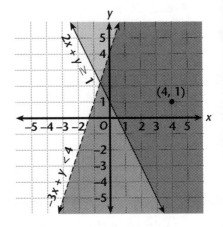

$2x + y \geq 1$	$-3x + y < 4$
$2(0) + 0 \overset{?}{\geq} 1$	$-3(0) + 0 \overset{?}{<} 4$
$0 \not\geq 1$	$0 < 4$
(0, 0) is not in the solution of $2x + y \geq 1$.	(0, 0) is in the solution of $-3x + y < 4$.

To check, pick a point in the region where both shaded portions overlap. See if the point makes both inequalities true.
For (4, 1):

$2x + y \geq 1$ $-3x + y < 4$
$2(4) + 1 \overset{?}{\geq} 1$ $-3(4) + 1 \overset{?}{<} 4$
$8 + 1 \overset{?}{\geq} 1$ $-12 + 1 \overset{?}{<} 4$
$9 \geq 1$ √ $-11 < 4$ √

Use graph paper to solve each system.

11. $\begin{cases} x - y > 5 \\ 2x + y > 4 \end{cases}$

12. $\begin{cases} 3x - 5y \leq 10 \\ 3y < -2x + 3 \end{cases}$

13. $\begin{cases} 3x - 4y \leq 12 \\ x - y \geq 1 \end{cases}$

14. $\begin{cases} 2x - y \geq 4 \\ 2y > x - 10 \end{cases}$

NAME _____ CLASS _____ DATE _____

 **Reteaching**
2.8 Classic Applications

◆ **Skill A** Representing unknown values in classic algebraic problems

Recall In a system of linear equations, if there are two unknowns, there must be two equations in which the two unknowns are represented.

◆ **Example 1**
Represent the following as a system of linear equations:
A gourmet shop owner mixes coffee worth $1.92 per pound with coffee worth $2.72 per pound to make a blend that will sell at $2.24 per pound. If she makes 30 pounds of the blend, how many pounds of each kind of coffee does she use?

◆ **Solution**
Define the variables. Let x represent the number of pounds of coffee worth $1.92 per pound.
Let y represent the number of pounds of coffee worth $2.72 per pound.
Write the equations.
$$x + y = 30 \qquad \text{Weight of mixture}$$
$$1.92x + 2.72y = 2.24(30) \qquad \text{Value of mixture}$$

◆ **Example 2**
Michael has 18 coins in his pocket, consisting of some quarters and dimes. The total value of all the coins is $3.45. Find the number of each kind of coin.

◆ **Solution**
Define the variables. Let x represent the number of quarters.
Let y represent the number of dimes.
Write the equations:
$$x + y = 18 \qquad \text{Number of coins}$$
$$0.25x + 0.10y = 3.45 \qquad \text{Value of coins}$$

◆ **Example 3**
The sum of the digits of a two-digit number is 13. The number is 4 more than five times the units digit. Find the number.

◆ **Solution**
Let x represent the tens digit.
Let y represent the ones digit.
$$x + y = 13 \qquad \text{Sum of digits}$$
$$10x + y = 5y + 4 \qquad \text{Number in relation to digits}$$

Represent each of the following as a system of linear equations:

1. How many pounds of candy worth $1.35 a pound should be mixed with candy worth $1.95 a pound to make 80 pounds of candy to sell at $1.50 a pound? _____

2. Jackie has $91 in $5 bills and $2 bills. He has 23 bills in all. How many $5 bills does he have? _____

3. The sum of the digits of a two-digit number is 9. The number is 6 times the ones digit. Find the number. _____

◆ **Skill B** Solving classic algebraic problems by using systems of linear equations

Recall The classic algebraic problems involve coins, mixtures, age, digits, and motion.

◆ **Example 1**
An airplane flies 720 miles with the wind in 2 hours. When flying against the wind, it takes 3 hours to fly the same distance. Find the rate of speed of the airplane and the rate at which the wind is blowing.

◆ **Solution**
Let x represent the rate of speed of the airplane.
Let y represent the rate of speed at which the wind is blowing.

rate × time = distance
$(x + y) \cdot 2 = 720$ With the wind
$(x - y) \cdot 3 = 720$ Against the wind

Solve the system by elimination.
$2x + 2y = 720$
$3x - 3y = 720$

Multiply the first equation by 3 and the second equation by 2. Then add the equations.

$6x + 6y = 2160$
$\underline{6x - 6y = 1440}$
$12x \quad\;\; = 3600$
$\quad\quad x = 300$

Substitute in the first equation to find y.

$2(300) + 2y = 720$
$\quad\quad\quad 2y = 120$
$\quad\quad\quad\; y = 60$

The airplane is flying at 300 mph and the wind speed is 60 mph.

Solve each problem by using a system of linear equations.

4. A rower can row 24 miles downstream (with the current) in 3 hours. The return trip over the same distance (against the current) takes 6 hours. Find his rowing rate in still water and the rate of the current.

_____

_____

5. Harriet is one-third as old as her sister, Jill. In 6 years she will be one-half as old as Jill. How old is Harriet now?

_____

_____

NAME _____ CLASS _____ DATE _____

 **Reteaching**
3.1 Using Matrices to Display Data

◆ **Skill A** Determining whether matrices are equal

Recall The dimension of a matrix is its number of rows and its number of columns, listed in that order. Two matrices are equal when their dimensions are the same and the corresponding entries are equal.

◆ **Example 1**
Does $M = N$?

$$M = \begin{bmatrix} 3^2 & \sqrt{64} & (5-2) \\ 4(-3) & \dfrac{30}{5} & 14 \end{bmatrix} \qquad N = \begin{bmatrix} \dfrac{36}{4} & (5+3) & \sqrt{9} \\ (-6)2 & 6 & (20-6) \end{bmatrix}$$

◆ **Solution**
The dimensions of the two matrices are equal because each matrix has 2 rows and 3 columns. Next, find the values of the entries at corresponding addresses in each matrix. For example, the value of m_{11} (row 1, column 1 of matrix M) is 3^2, or 9. At the corresponding address of matrix N, n_{11}, is $\dfrac{36}{4}$, or 9.

$\begin{aligned} m_{12} &= \sqrt{64} = 8 \\ m_{13} &= 5 - 2 = 3 \\ m_{21} &= 4(-3) = -12 \\ m_{22} &= \dfrac{30}{5} = 6 \\ m_{23} &= 14 \end{aligned}$ $\begin{aligned} n_{12} &= 5 + 3 = 8 \\ n_{13} &= \sqrt{9} = 3 \\ n_{21} &= (-6)2 = -12 \\ n_{22} &= 6 \\ n_{23} &= 20 - 6 = 14 \end{aligned}$

Both matrices simplify to $\begin{bmatrix} 9 & 8 & 3 \\ -12 & 6 & 14 \end{bmatrix}$.

Therefore $M = N$.

◆ **Example 2**
Is $R = S$?

$$R = \begin{bmatrix} -5 & 2 & 0 & -1 \\ 4 & -3 & 6 & -7 \end{bmatrix} \qquad S = \begin{bmatrix} -5 & 4 \\ 2 & -3 \\ 0 & 6 \\ -1 & -7 \end{bmatrix}$$

◆ **Solution**
The dimension of matrix R is 2×4 because it has 2 rows and 4 columns. The dimension of matrix S is 4×2 because it has 4 rows and 2 columns. Therefore, even though the values of the entries are the same, $R \neq S$.

Use the given matrices to answer each question.

$$A = \begin{bmatrix} 2 & 0 & -3 \\ -1 & 4 & 6 \\ 3 & -2 & 5 \end{bmatrix} \quad B = \begin{bmatrix} \dfrac{6}{3} & 0 & -\dfrac{24}{8} \\ (-1)1 & \sqrt{16} & 3(2) \\ (8-5) & 0-2 & \sqrt{25} \end{bmatrix} \quad C = \begin{bmatrix} 0 & -3 \\ 4 & 6 \\ -2 & 5 \end{bmatrix} \quad D = \begin{bmatrix} 2 & 0 & -3 \\ -1 & 4 & 6 \end{bmatrix}$$

1. What are the dimensions of each matrix? _____

2. Does entry $a_{22} = b_{22}$? Explain. _____

---

3. Is $A = B$? Why? _____

4. Is $C = D$? Why? _____

◆ **Skill B**   Finding unknown values in equal matrices

Recall   Corresponding entries in equal matrices are equal.

◆ **Example**
Matrices C and D are equal. Find the values for l, m, and n.

$$C = \begin{bmatrix} -3m + 6 & -4l \\ m + 12 & -4m \end{bmatrix} \qquad D = \begin{bmatrix} 12 & 2n \\ -4l + 2 & 2n \end{bmatrix}$$

◆ **Solution**
Since matrices C and D are equal, the corresponding entries are equal. Write equations that express the equalities.

$$-3m + 6 = 12 \qquad\qquad -4l = 2n$$
$$m + 12 = -4l + 2 \qquad\qquad -4m = 2n$$

Solve the equation that has only one variable first. Then that value can be substituted in other equations.

$$-3m + 6 = 12 \qquad\qquad -4m = 2n$$
$$-3m = 6 \qquad\qquad -4(-2) = 2n \qquad \text{Substitute } -2 \text{ for } m.$$
$$m = -2 \qquad\qquad 8 = 2n$$
$$4 = n \qquad \text{Simplify.}$$

$$-4l = 2n$$
$$-4l = 2(4) \qquad \text{Substitute 4 for } n.$$
$$-4l = 8$$
$$l = -2 \qquad \text{Simplify.}$$
The values are $l = -2$, $m = -2$, and $n = 4$.

Assuming each pair of matrices is equal, find the unknown values.

5. $R = \begin{bmatrix} 5x + 2 & -4y + 3 \\ 3y & 2xy \end{bmatrix}$ $\qquad S = \begin{bmatrix} 3z - 6 & 15 \\ z + 15 & -4z \end{bmatrix}$

_____

6. $T = \begin{bmatrix} 3a + 2 & ab - 5 & -3c + 5 \\ 4b - 5 & 2d - 3 & ae - 4 \end{bmatrix}$ $\qquad U = \begin{bmatrix} d + 1 & e & 2bd - 4 \\ 7 & b + 2 & \frac{bd}{-2} \end{bmatrix}$

_____

7. How can you check to see whether the values are correct?

_____

_____

Reteaching
3.2 Adding and Subtracting Matrices

◆**Skill A** Subtracting by adding opposites

Recall To subtract, change the second term to its opposite and add it to the first term.

◆ **Example**
Subtract: **a.** $12 - 17$ **b.** $-14 - (-32)$

◆ **Solution**

a. first term		second term	b. first term		second term
12	$-$	17	-14	$-$	(-32)
12	$+$	(-17)	-14	$+$	32
	-5			18	

Perform each indicated operation.

1. $5 - 11$ _____

2. $-13 - 12$ _____

3. $15 - (-32)$ _____

4. $24 - 17$ _____

5. $-44 - 48$ _____

6. $37 - (-15)$ _____

7. $55 - (-38)$ _____

8. $-42 - (-63)$ _____

◆**Skill B** Adding matrices

Recall Matrices can be added only if they have the same dimensions. If they do, then add corresponding entries.

◆ **Example**

If $A = \begin{bmatrix} -5 & 3 \\ \frac{5}{2} & -1 \end{bmatrix}$ and $B = \begin{bmatrix} 2.6 & -4.5 \\ -2.5 & 6 \end{bmatrix}$, find $A + B$.

◆ **Solution**
Since the matrices have the same dimensions, corresponding entries can be added.

$$A + B = \begin{bmatrix} -5 + 2.6 & 3 + (-4.5) \\ \frac{5}{2} + (-2.5) & -1 + 6 \end{bmatrix} = \begin{bmatrix} -2.4 & -1.5 \\ 0 & 5 \end{bmatrix}$$

Use the following matrices to find each indicated sum, if possible:

$$A = \begin{bmatrix} 3.2 & -5.4 & 8 \\ 2.7 & 1.5 & -3 \end{bmatrix} \qquad B = \begin{bmatrix} \frac{3}{2} & -\frac{5}{2} & 7 \\ -\frac{7}{2} & -\frac{3}{2} & 4 \end{bmatrix} \qquad C = \begin{bmatrix} 4 & 0 \\ -2 & 6 \\ 0 & -1 \end{bmatrix} \qquad D = \begin{bmatrix} -\frac{3}{2} & \frac{7}{2} \\ \frac{5}{2} & \frac{3}{2} \\ 7 & -4 \end{bmatrix}$$

9. $A + B$ _____

10. $B + C$ _____

11. $C + D$ _____

◆ Skill C Subtracting matrices by adding opposites

Recall The opposite of a matrix is formed by changing the sign of each entry in the matrix. The sum of a matrix and its opposite is the identity, or zero matrix.

◆ **Example 1**
Find the opposite of matrix A.

$$A = \begin{bmatrix} \frac{3}{5} & -1 & 0 \\ 2.6 & 4 & -5 \end{bmatrix}$$

◆ **Solution**

The opposite of A is $\begin{bmatrix} -\frac{3}{5} & 1 & 0 \\ -2.6 & -4 & 5 \end{bmatrix}$ since

$$\begin{bmatrix} \frac{3}{5} & -1 & 0 \\ 2.6 & 4 & -5 \end{bmatrix} + \begin{bmatrix} -\frac{3}{5} & 1 & 0 \\ -2.6 & -4 & 5 \end{bmatrix} = \begin{bmatrix} 0 & 0 & 0 \\ 0 & 0 & 0 \end{bmatrix}.$$

◆ **Example 2**

If $A = \begin{bmatrix} 3 & -5 \\ 2 & 6 \\ -4 & 0 \end{bmatrix}$ and $B = \begin{bmatrix} -4 & 7 \\ 2 & -6 \\ 5 & -3 \end{bmatrix}$, find $A - B$.

◆ **Solution**
First find the opposite of B.

$$B = \begin{bmatrix} -4 & 7 \\ 2 & -6 \\ 5 & -3 \end{bmatrix} \qquad \text{Opposite of } B = \begin{bmatrix} 4 & -7 \\ -2 & 6 \\ -5 & 3 \end{bmatrix}$$

Then add the opposite of B to A.

$$\begin{bmatrix} 3 & -5 \\ 2 & 6 \\ -4 & 0 \end{bmatrix} + \begin{bmatrix} 4 & -7 \\ -2 & 6 \\ -5 & 3 \end{bmatrix} = \begin{bmatrix} 7 & -12 \\ 0 & 12 \\ -9 & 3 \end{bmatrix}$$

Perform each indicated subtraction.

12. $\begin{bmatrix} 5 & -4 & 2 \\ -1 & 3 & -6 \end{bmatrix} - \begin{bmatrix} -3 & -2 & 6 \\ 4 & -5 & 0 \end{bmatrix}$ _____

13. $\begin{bmatrix} 5 & 9 \\ -3 & 2 \end{bmatrix} - \begin{bmatrix} 3 & -5 \\ -9 & 4 \end{bmatrix}$ _____

14. $\begin{bmatrix} 3 & -4 \\ 0 & 2 \\ 8 & -5 \end{bmatrix} - \begin{bmatrix} -6 & 3 \\ -7 & 5 \\ -2 & 6 \end{bmatrix}$ _____

Reteaching

3.3 Exploring Matrix Multiplication

◆ **Skill A** Determining whether two matrices can be multiplied

Recall To multiply two matrices, the column dimension of the first must equal the row dimension of the second.

◆ **Example 1**

Can matrix A be multiplied by matrix B?

$$A = \begin{bmatrix} 2 & 5 & -1 \\ -3 & 2 & 6 \end{bmatrix} \qquad B = \begin{bmatrix} 5 & -1 & 4 \\ 2 & 6 & 3 \\ 4 & 0 & -5 \end{bmatrix}$$

◆ **Solution**

Determine the dimensions of each matrix. The dimensions of matrix A are 2×3. The dimensions of matrix B are 3×3.

rows		columns		rows		columns
2	×	③		③	×	3

Since the column dimension of A is the same as the row dimension of B, matrix A can be multiplied by matrix B.

◆ **Example 2**

If the dimensions of matrix C are 2×3 and the dimensions of matrix D are 2×4, can matrix C be multiplied by matrix D?

◆ **Solution**

Compare the column number of the first with the row number of the second.

$$2 \times ③ \qquad ② \times 4$$

Since these two numbers are not equal, matrices C and D cannot be multiplied.

State whether each of the following pairs of matrices can be multiplied:

1. $A: 3 \times 5,\ B: 5 \times 4$ _____

2. $C: 4 \times 3,\ D: 3 \times 4$ _____

3. $R: 3 \times 3,\ S: 2 \times 3$ _____

4. $M = \begin{bmatrix} 5 & 1 & 3 & 6 \\ -1 & 2 & -4 & 5 \end{bmatrix} \quad N = \begin{bmatrix} 5 & -2 & 1 & 6 \\ -1 & 4 & 3 & -4 \\ -6 & -3 & 2 & 4 \end{bmatrix}$ _____

5. $P = \begin{bmatrix} 2 & 3 & -1 \\ 4 & -2 & 0 \end{bmatrix} \quad Q = \begin{bmatrix} 2 & -2 & 4 \\ 1 & 3 & 6 \end{bmatrix}$ _____

6. $T = \begin{bmatrix} 1 & 0 & -3 \\ 2 & -1 & 4 \\ 5 & -4 & 3 \end{bmatrix} \quad U = \begin{bmatrix} -4 & 2 \\ 0 & 5 \\ -1 & 4 \end{bmatrix}$ _____

◆ **Skill B**   Multiplying matrices

Recall   The product of numbers with like signs is positive. The product of numbers with unlike signs is negative. Multiplication of matrices involves each row of the first matrix and each column of the second matrix.

◆ **Example**

Multiply $\begin{bmatrix} 2 & -3 \\ 4 & 5 \end{bmatrix}\begin{bmatrix} 1 & 5 \\ 0 & -2 \end{bmatrix}$, if possible.

◆ **Solution**

Since the dimensions of the first matrix are 2×2 and the dimensions of the second matrix are 2×2, multiplication is possible.

Column
1   2

Since there are 2 rows in the first matrix and 2 columns in the second matrix, there are 4 row-column combinations.

$\begin{matrix} \textbf{Row} \ \mathbf{1} \\ \mathbf{2} \end{matrix}\begin{bmatrix} 2 & -3 \\ 4 & 5 \end{bmatrix}\begin{bmatrix} 1 & 5 \\ 0 & -2 \end{bmatrix}$

Column 1:

Multiply the first entry in row 1 by the first entry in column 1. Multiply the second entry in row 1 by the second entry in column 1. Add the resulting products.

$2 \cdot 1 + (-3) \cdot 0$
$2 + 0 = 2$

Column 2:

Repeat this procedure for each row-column combination.

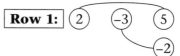

$2 \cdot 5 + (-3) \cdot (-2)$
$10 + 6 = 16$

Column 1:

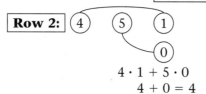

$4 \cdot 1 + 5 \cdot 0$
$4 + 0 = 4$

Column 2:

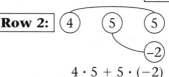

$4 \cdot 5 + 5 \cdot (-2)$
$20 + (-10) = 10$

The product matrix has four entries. The address in the product matrix is found by using the row-column numbers which produced the entry.

Row 1, Column 1 $\longrightarrow$ $\begin{bmatrix} 2 & 16 \\ 4 & 10 \end{bmatrix}$ $\longleftarrow$ Row 1, Column 2
Row 2, Column 1 $\longrightarrow$ $\longleftarrow$ Row 2, Column 2

Find each product, if possible.

7. $\begin{bmatrix} -4 & 0 \\ 2 & -1 \end{bmatrix}\begin{bmatrix} 4 & 6 \\ -2 & -3 \end{bmatrix}$ _____

8. $\begin{bmatrix} 5 & 1 \\ 0 & 2 \end{bmatrix}\begin{bmatrix} 3 \\ -4 \end{bmatrix}$ _____

9. $\begin{bmatrix} 5 & 4 \\ -3 & 6 \\ 2 & 5 \end{bmatrix}\begin{bmatrix} 6 & 1 & 0 \\ -2 & 5 & 1 \\ 4 & 3 & -4 \end{bmatrix}$ _____

10. $\begin{bmatrix} 1 & 4 \end{bmatrix}\begin{bmatrix} 2 \\ 3 \end{bmatrix}$

_____

11. $\begin{bmatrix} 4 & 3 & 2 \\ -1 & 0 & -5 \end{bmatrix}\begin{bmatrix} 5 & -2 \\ -1 & 3 \\ 0 & -4 \end{bmatrix}$ _____

12. $\begin{bmatrix} 1 \\ 2 \\ 3 \end{bmatrix}\begin{bmatrix} -3 & -2 & -1 \end{bmatrix}$

_____

Reteaching
3.4 Multiplicative Inverse of a Matrix

◆ **Skill A**   Writing a system of equations to find an inverse matrix

Recall   The product of a number and its multiplicative inverse is 1, the multiplicative identity.

If $a \cdot \frac{1}{a} = 1$, then $\frac{1}{a}$ is the multiplicative inverse of a.

The identity for a 2×2 square matrix is $\begin{bmatrix} 1 & 0 \\ 0 & 1 \end{bmatrix}$.

◆ **Example**
Write two systems of linear equations that can be used to find the inverse of

$A = \begin{bmatrix} 5 & -2 \\ -3 & 1 \end{bmatrix}$.

◆ **Solution**
Represent the inverse of A or A^{-1} as $\begin{bmatrix} a & b \\ c & d \end{bmatrix}$.

Then: $\begin{bmatrix} 5 & -2 \\ -3 & 1 \end{bmatrix}\begin{bmatrix} a & b \\ c & d \end{bmatrix} = \begin{bmatrix} 1 & 0 \\ 0 & 1 \end{bmatrix}$ A matrix times its inverse produces the identity.

$\begin{bmatrix} 5a - 2c & 5b - 2d \\ -3a + c & -3b + d \end{bmatrix} = \begin{bmatrix} 1 & 0 \\ 0 & 1 \end{bmatrix}$ Perform matrix multiplication on the left side.

Use matrix equality to pair the equations with the same variables.

$$\begin{cases} 5a - 2c = 1 \\ -3a + c = 0 \end{cases} \qquad \begin{cases} 5b - 2d = 0 \\ -3b + d = 1 \end{cases}$$

Write two systems of linear equations that can be used to find the inverse of each matrix.

1. $A = \begin{bmatrix} 4 & 6 \\ -2 & 5 \end{bmatrix}$ _____

2. $B = \begin{bmatrix} 2 & -6 \\ 1 & 3 \end{bmatrix}$ _____

_____

_____

3. $C = \begin{bmatrix} 5 & 9 \\ -4 & -2 \end{bmatrix}$ _____

4. $D = \begin{bmatrix} -1 & 0 \\ 5 & -3 \end{bmatrix}$ _____

_____

_____

5. $E = \begin{bmatrix} 1 & -2 \\ 4 & 6 \end{bmatrix}$ _____

6. $F = \begin{bmatrix} 9 & -5 \\ -2 & 4 \end{bmatrix}$ _____

_____

_____

7. $G = \begin{bmatrix} -1 & 4 \\ -2 & 5 \end{bmatrix}$ _____

8. $H = \begin{bmatrix} 5 & -1 \\ 0 & -2 \end{bmatrix}$ _____

_____

_____

◆ **Skill B** Determining the inverses of matrices

Recall Systems of linear equations in two variables can be solved by graphing, substitution, or elimination.

◆ **Example**
Find the inverse of $A = \begin{bmatrix} 5 & 2 \\ 7 & 3 \end{bmatrix}$.

◆ **Solution**
Represent A^{-1} as $\begin{bmatrix} a & b \\ c & d \end{bmatrix}$. Use it to write two systems of linear equations.

$$\begin{bmatrix} 5 & 2 \\ 7 & 3 \end{bmatrix}\begin{bmatrix} a & b \\ c & d \end{bmatrix} = \begin{bmatrix} 1 & 0 \\ 0 & 1 \end{bmatrix}$$

$$\begin{bmatrix} 5a + 2c & 5b + 2d \\ 7a + 3c & 7b + 3d \end{bmatrix} = \begin{bmatrix} 1 & 0 \\ 0 & 1 \end{bmatrix}$$

$$\begin{cases} 5a + 2c = 1 \\ 7a + 3c = 0 \end{cases} \quad \begin{cases} 5b + 2d = 0 \\ 7b + 3d = 1 \end{cases}$$

Choose a method for solving the systems of equations. In this case, elimination is convenient. Use the first system to solve for a and c. Multiply the first equation by 3 and the second equation by -2.

$$\begin{array}{rcl} 5a + 2c = 1 & \longrightarrow & 15a + 6c = 3 \\ 7a + 3c = 0 & \longrightarrow & \underline{-14a - 6c = 0} \\ & & a \qquad = 3 \end{array}$$

Substitute 3 for a in either of the original equations, and solve for c.

$$\begin{array}{ll} 5(3) + 2c = 1 & 7(3) + 3c = 0 \\ 15 + 2c = 1 & 21 + 3c = 0 \\ 2c = -14 & 3c = -21 \\ c = -7 & c = -7 \end{array}$$

Use the second system of equations to solve for b and d. Multiply the first equation by 3 and the second equation by -2.

$$\begin{array}{rcl} 5b + 2d = 0 & \longrightarrow & 15b + 6d = 0 \\ 7b + 3d = 1 & \longrightarrow & \underline{-14b - 6d = -2} \\ & & b \qquad = -2 \end{array}$$

$$\begin{array}{l} 5(-2) + 2d = 0 \\ -10 + 2d = 0 \\ 2d = 10 \\ d = 5 \end{array}$$

Therefore, $A^{-1} = \begin{bmatrix} 3 & -2 \\ -7 & 5 \end{bmatrix}$

Find the inverse of each matrix.

9. $\begin{bmatrix} 3 & 4 \\ 2 & 3 \end{bmatrix}$ _____

10. $\begin{bmatrix} 3 & 5 \\ 1 & 2 \end{bmatrix}$ _____

11. $\begin{bmatrix} 7 & 17 \\ 2 & 5 \end{bmatrix}$ _____

12. $\begin{bmatrix} 7 & 6 \\ 8 & 7 \end{bmatrix}$ _____

Reteaching
3.5 Solving Matrix Equations

◆**Skill A** Writing a system of linear equations in matrix form

Recall A system of linear equations can be expressed in the form $AX = B$, where A is the coefficient matrix, X is the variable matrix, and B is the constant matrix.

◆ **Example 1**
Write the following system of linear equations as a matrix equation:
$$\begin{cases} 2x + 5y = 14 \\ 3x + 4y = 7 \end{cases}$$

◆ **Solution**
Form a 2×2 matrix with the coefficients of x and y.

$$A = \begin{bmatrix} 2 & 5 \\ 3 & 4 \end{bmatrix}$$

Form a 2×1 matrix with the variables.

$$X = \begin{bmatrix} x \\ y \end{bmatrix}$$

Form a 2×1 matrix with the constants.

$$B = \begin{bmatrix} 14 \\ 7 \end{bmatrix}$$

Write the equation in the form $AX = B$.

$$\begin{bmatrix} 2 & 5 \\ 3 & 4 \end{bmatrix} \begin{bmatrix} x \\ y \end{bmatrix} = \begin{bmatrix} 14 \\ 7 \end{bmatrix}$$

◆ **Example 2**
Write the following system of linear equations as a matrix equation:
$$\begin{cases} -5x + 2y = 3 \\ y = 3x - 1 \end{cases}$$

◆ **Solution**
Rewrite the second equation in standard form.

$$\begin{aligned} y &= 3x - 1 \\ y - 3x &= -1 \\ -3x + y &= -1 \qquad \text{Commutative Property of Addition} \end{aligned}$$

Then form the coefficient, variable, and constant matrices.

$$\begin{cases} -5x + 2y = 3 \\ -3x + y = -1 \end{cases} \qquad \begin{bmatrix} -5 & 2 \\ -3 & 1 \end{bmatrix} \begin{bmatrix} x \\ y \end{bmatrix} = \begin{bmatrix} 3 \\ -1 \end{bmatrix}$$

Express each system of linear equations in matrix form.

1. $\begin{cases} 8x + 3y = 4 \\ -3x + 2y = 11 \end{cases}$ _____

2. $\begin{cases} 2x - 5y = 16 \\ 7x + 9y = 3 \end{cases}$ _____

3. $\begin{cases} 9x + 6y = 15 \\ 3x + 2y = 5 \end{cases}$ _____

4. $\begin{cases} -5x + 2y = -3 \\ 10x + -4y = 6 \end{cases}$ _____

5. $\begin{cases} 2y = 5x + 3 \\ 3x - y = 1 \end{cases}$ _____

6. $\begin{cases} 9x = 4 + 5y \\ 3y = 2x + 1 \end{cases}$ _____

◆**Skill B** Solving systems of equations by using matrices

Recall The product of a matrix and its inverse is the identity matrix.
$$AA^{-1} = I$$

◆ **Example**

Solve the system of linear equations using matrices.
$$\begin{cases} 5x - 2y = -3 \\ 3x - y = 1 \end{cases}$$

◆ **Solution**

Write the system in matrix form.
$$\begin{cases} 5x - 2y = -3 \\ 3x - y = 1 \end{cases} \longrightarrow \begin{bmatrix} 5 & -2 \\ 3 & -1 \end{bmatrix}\begin{bmatrix} x \\ y \end{bmatrix} = \begin{bmatrix} -3 \\ 1 \end{bmatrix}$$

Find the inverse of the coefficient matrix.

Use $\begin{bmatrix} 5 & -2 \\ 3 & -1 \end{bmatrix}\begin{bmatrix} a & b \\ c & d \end{bmatrix} = \begin{bmatrix} 1 & 0 \\ 0 & 1 \end{bmatrix}$ to form two systems of linear equations, and solve for the unknowns a, b, c, and d.

$$\begin{cases} 5a - 2c = 1 \\ 3a - c = 0 \end{cases} \qquad \begin{cases} 5b - 2d = 0 \\ 3b - d = 1 \end{cases}$$

Using elimination:

$5a - 2c = 1$	$5b - 2d = 0$
$\underline{-6a + 2c = 0}$	$\underline{-6b + 2d = -2}$
$-a \qquad = 1$	$-b \qquad = -2$
$a = -1$	$b = 2$

$3(-1) - c = 0$	$5(2) - 2d = 0$
$-3 - c = 0$	$10 - 2d = 0$
$-c = 3$	$-2d = -10$
$c = -3$	$d = 5$

The inverse of the matrix $\begin{bmatrix} 5 & -2 \\ 3 & -1 \end{bmatrix}$ is $\begin{bmatrix} -1 & 2 \\ -3 & 5 \end{bmatrix}$. Multiply each side of the original matrix equation by the inverse.

$$\begin{bmatrix} -1 & 2 \\ -3 & 5 \end{bmatrix}\begin{bmatrix} 5 & -2 \\ 3 & -1 \end{bmatrix}\begin{bmatrix} x \\ y \end{bmatrix} = \begin{bmatrix} -1 & 2 \\ -3 & 5 \end{bmatrix}\begin{bmatrix} -3 \\ 1 \end{bmatrix}$$

$$\begin{bmatrix} 1 & 0 \\ 0 & 1 \end{bmatrix}\begin{bmatrix} x \\ y \end{bmatrix} = \begin{bmatrix} 3 + 2 \\ 9 + 5 \end{bmatrix}$$ Coefficient matrix times its inverse equals identity.

$$\begin{bmatrix} x \\ y \end{bmatrix} = \begin{bmatrix} 5 \\ 14 \end{bmatrix}$$ Identity matrix times variable matrix equals variable matrix.

The solution is $x = 5$ and $y = 14$.

Solve each system by using matrices.

7. $\begin{cases} 7x + 4y = 26 \\ 5x + 3y = 19 \end{cases}$ _____

8. $\begin{cases} 5x + 3y = -7 \\ 8x + 5y = -11 \end{cases}$ _____

9. $\begin{cases} 2x + 3y = 96 \\ y - x = 2 \end{cases}$ _____

10. $\begin{cases} 9x + 2y = 2 \\ 21x + 6y = 4 \end{cases}$ _____

NAME _____ CLASS _____ DATE _____

Reteaching
4.1 Experimental Probability

◆ **Skill A**   Writing fractions as percents

Recall   To change a fraction to a percent, first divide the numerator of the fraction by its denominator to change it to a decimal. Then move the decimal point two places to the right and include a percent sign.

◆ **Example**
Express each fraction as a percent.

    a. $\frac{3}{8}$ **b.** $\frac{5}{9}$

◆ **Solution**
    a.   Divide the numerator, 3, by the denominator, 8.

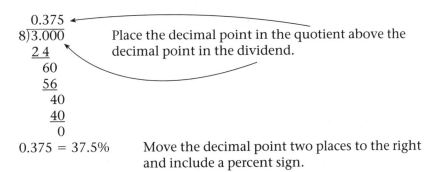

        $0.375 = 37.5\%$ Move the decimal point two places to the right and include a percent sign.

    b.   Divide the numerator, 5, by the denominator, 9.

```
  0.555 ...      Use three dots in the quotient to indicate that
9)5.000          the number pattern repeats, or put a bar over
  4 5            the repeating number pattern.
   50            0.555 ... = 0.55̄5
   45
   50
   45
    5
```
$0.55\overline{5} \approx 55.5\%$

Express each fraction as a decimal and as a percent.

1. $\frac{4}{5}$ _____   2. $\frac{5}{8}$ _____

3. $\frac{3}{2}$ _____   4. $\frac{2}{7}$ _____

5. $\frac{1}{3}$ _____   6. $\frac{4}{15}$ _____

HRW Algebra One Interactions Course 2   **Reteaching 4.1**   **39**

◆ **Skill B** Determining experimental probabilities

Recall The experimental probability, P, of an event is equal to

$$\frac{\text{the number of times a successful event occurs}}{\text{the number of trials in the experiment}} = \frac{f}{t}.$$

◆ **Example 1**

The results of rolling a number cube 50 times are given in the table. Find the experimental probability of rolling a 3.

Number	1	2	3	4	5	6
Frequency	7	11	8	7	4	13

◆ **Solution**

The number 3 was rolled 8 times in the 50 trials.

$$P = \frac{\text{number of successes}}{\text{number of trials}} \qquad P = \frac{8}{50} = \frac{4}{25}, \text{ or } 16\%$$

The experimental probability of rolling a 3 is $\frac{4}{25}$, or 16%.

◆ **Example 2**

Use the table in Example 1 to find the experimental probability for rolling an even number.

◆ **Solution**

The even numbers are 2, 4 and 6. The number 2 was rolled 11 times, 4 was rolled 7 times, and 6 was rolled 13 times. There were $11 + 7 + 13$, or 31, successes in 50 trials. The experimental probability of rolling an even number is $\frac{31}{50}$, or 62%.

The results of spinning a spinner with red, blue, green, and yellow regions 100 times are given in the table.

Region	red	blue	green	yellow
Frequency	17	21	19	43

7. Find the experimental probability of landing in the

   red region. _____

8. Find the experimental probability of landing in the

   green region. _____

9. Find the experimental probability of landing in either the blue or

   yellow region. _____

10. Find the experimental probability of not landing in the

    red region. _____

11. Find the experimental probability of landing in the red, blue, green,

    or yellow region. _____

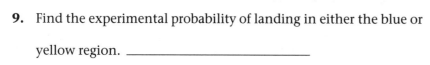

Reteaching
4.2 Exploring Simulations

◆ **Skill A**   Using coins as simulation devices

Recall   Because a coin toss has only two possible outcomes—heads or tails—it is the random generator to use when the situation being simulated has exactly two outcomes.

◆ **Example**
Chris and Paul play checkers every day. Before starting, Paul draws without looking from a box containing one red piece and one black piece to decide which color he will play. In the month of September, how many days is Paul likely to choose red? What is the experimental probability?

◆ **Solution**
Since there are exactly two outcomes, red and black, a coin can be used to simulate the outcomes.

Let heads represent red and tails represent black. Let each toss represent a day's draw. Because the problem specifies the month of September, which has 30 days, 30 trials should be generated and the results recorded.

Sample set of outcomes:

THTHTTTHTH
HHHTHHTTTH
THHTTTHHHT

In this simulation, there are 15 heads. This represents 15 days in September that Paul is likely to choose red. Thus, the experimental probability of Paul choosing red is $\frac{15}{30} = \frac{1}{2}$, or 50%.

Using a coin as a random generator, design and perform a simulation for each situation described.

1. A teacher asked 24 students to guess the answer to this question: "I have a quarter in my pocket–true or false?" If the teacher had no quarter, how many students guessed correctly?

_____

_____

2. As part of her weekly workout, Arlene does sit-ups and push-ups. To decide which she should do first, she looks at her digital watch. If the seconds indicator is an even number, she does sit-ups first. What is the probability that she starts her workout with sit-ups?

_____

_____

◆ **Skill B**   Designing simulations

Recall   Random values can be generated by rolling a die, tossing a coin, spinning a spinner, or using a calculator.

◆ **Example**

Design a simulation to find the experimental probability for the following situation:

> The weather forecaster predicts a 50% chance of rain on Saturday and an 80% chance of rain on Sunday. What is the experimental probability that it will rain on both days?

◆ **Solution**

Decide what type of random number generator to use for the simulation. Number cubes of different colors could be used for both predictions. The numbers 1, 2, and 3 on one cube could represent rain while 4, 5, and 6 could represent no rain on Saturday. On the second cube, the numbers 1, 2, 3, and 4 could represent rain on Sunday, while 5 would represent no rain. A roll of 6 would mean roll again. Or a coin could be flipped for Saturday's results, using heads for rain and tails for no rain.

Each pair of results would represent one trial, or outcome. Generate 20 trials and record the results.

Design a simulation to find the experimental probability for each situation described.

3.   If a playing card is drawn at random from a standard 52-card deck,

what is the probability that it will be a red card? _____

_____

4.   A candy dish contains 60 candies, 10 each of six colors: red, green, yellow, blue, tan, and dark brown. If Paula takes one candy from the dish without looking, what is the probability that she will pick a

tan candy? _____

_____

5.   A 40% chance of rain on Saturday and a 50% chance of rain on Sunday is predicted. What is the probability that it will not rain on either day?

_____

_____

Reteaching
4.3 Extending Statistics

◆ **Skill A** Finding the mean, median, and mode for sets of data

Recall The mean is the average for a set of numbers. To find the mean, add the set of numbers and divide by the number of terms in the set.

The median is the middle number for a set of numbers. To find the median, arrange the numbers in order. If there is an odd number of terms, use the middle term. If there is an even number of terms, use the mean of the two middle numbers.

The mode is the number that occurs most often in a set of numbers.

◆ **Example**
As a member of a bowling league, Ray bowled the following scores:
132, 185, 163, 191, 183, 166, 163, 190, 185, 163, 158, 190
Find the mean, median, and mode of Ray's bowling scores.

◆ **Solution**
Add the 12 numbers and divide the sum by 12.
132 + 185 + 163 + 191 + 183 + 166 + 163 + 190 + 185 + 163 + 158 + 190 = 2069

$\frac{2069}{12} \approx 172.42$ Round 172.42 to the nearest tenth.

The mean is approximately 172.4.

Arrange the scores from largest to smallest or smallest to largest. Since there are 12 scores, the median is the mean of the sixth and seventh terms.
132, 158, 163, 163, 163, 166, 183, 185, 185, 190, 190, 191
                                    \   /
                                  6th 7th

$\frac{166 + 183}{2} = \frac{349}{2} = 174.5$ Add the two terms and divide by 2.

The median is 174.5.
The number 163 occurs three times; therefore, the mode is 163.

1. Find the mean, median, and mode of the following data set:
    12, 21, 35, 35, 48, 56, 65, 74, 89

_____

2. The 15 students in a math class received the following scores:
    87, 64, 62, 95, 94, 55, 96, 43, 89, 94, 65, 92, 95, 95, 67
    Determine the mean, median, and mode of the scores.

_____

◆ **Skill B**   Constructing scatter plots

Recall   A scatter plot is a graph consisting of distinct points that represent data pairs.

A scatter plot is a visual way of determining whether two quantities are related.

◆ **Example**
The height and arm span, measured in centimeters, of 10 students, are given in the table. Construct a scatter plot of the data.

Height	153	165	177	149	156	139	143	170	173	166
Arm span	149	162	178	158	150	136	135	164	173	170

◆ **Solution**
Plot each data pair on a coordinate grid. Since the height data ranges from 139 to 177 and the arm span data ranges from 135 to 178, use the numbers from 130 to 180 on each axis. Use line breaks on each axis near the origin to indicate that the values between 0 and 130 are not shown.

Notice that the points tend to lie along a line, indicating a positive correlation between height and arm span.

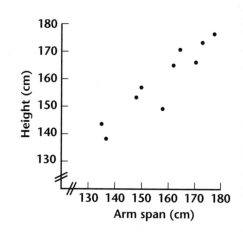

Use graph paper to construct a scatter plot for each situation.

3.   Mrs. Baker polled her class to obtain data concerning the length of time that students studied for a math test and their resulting grade.

Minutes studying	85	120	20	35	10	30	60	40	90	85	105	30	50	20	15	80	100	110	25	40	55	20	35	45	90
Test grade	90	95	30	50	42	75	80	55	93	78	88	40	75	62	37	75	92	90	53	75	79	71	73	79	84

4.   A class took a survey to compare the number of minutes spent on daily exercise with the number of sick days taken during a year.

Minutes of daily exercise	20	60	45	60	30	0	10	15	25	50	0	90	40	50	30	20	5	15	45	60
Number of sick days	3	1	1	0	2	10	3	2	3	4	15	0	2	1	3	4	8	5	1	1

Reteaching
4.4 Exploring the Addition Principle of Counting

◆ **Skill A**   Drawing and using Venn diagrams

Recall   In a Venn diagram, overlapping regions represent the intersection of two sets, or the word AND. Combined regions represent the union of two sets, or the word OR. The area outside a region represents the word NOT.

◆ **Example**
In a small Ghanaian village there are 38 children between 6 and 12 years old. Eighteen of the children eat kenkey and 27 eat fufu. This includes 7 children who eat both. Use a Venn diagram to determine how many children eat only fufu.

◆ **Solution**
Because some children eat both foods, draw two circles that overlap. In the part where the two circles overlap, place the number that represents the number of children who eat both foods.

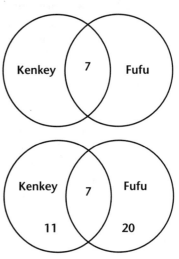

The total number of children who eat kenkey must be 18. Since 7 children that eat kenkey are already represented, place an additional 11 in the other part of the kenkey circle. The total number of children who eat fufu is 27. Since 7 that eat fufu are already represented, an additional 20 must be placed within the other part of the fufu circle. This region represents the 20 children who eat only fufu.

Represent each situation with a Venn diagram.

1. Fifteen children at a birthday party are taken to an ice-cream parlor. Seven order chocolate and 12 order vanilla. This includes 4 who order a combination of chocolate and vanilla.

2. At a summer camp, 36 campers were asked whether they play golf or tennis. Fifteen said they play golf, while 17 reported that they play tennis. This included 2 who play both and 6 who play neither.

◆ **Skill B**   Using the addition principle of counting

Recall   The total number of ways to choose A OR B = (number of ways to chose A + number of ways to choose B) − (number of ways to choose both A AND B).

◆ **Example 1**

A survey of student preferences for spring sports produced the following results:

	Baseball	Lacrosse	Total
Blue	14	16	30
Orange	12	19	31
Total	26	35	61

   a.   How many students surveyed were in the Orange group?
   b.   How many students preferred baseball?
   c.   How many students were in the Orange group AND preferred baseball?
   d.   How many students were in the Orange group OR preferred baseball?

◆ **Solution**

   a.   The total for the second row, Orange, is 31.
   b.   The total for the first column, Baseball, is 26.
   c.   The intersection of the first column, baseball, and the second row, Orange team is 12.
   d.   Number of students in Orange group OR preferred baseball =

	Number of students in Orange group	31
+	Number of students preferring baseball	+ 26
		57
−	Number of students in Orange group AND preferring baseball	− 12
		45 students

3.   The Parents' Sports Club conducted a survey to determine whether to hold an auction or run a 50-50 to raise funds. The table shows the results of the survey.

	Auction	50-50	Total
Men	44	91	135
Women	116	35	151
Total	160	126	286

   a.   How many parents surveyed were men? _____

   b.   How many parents preferred a 50-50? _____

   c.   How many parents surveyed were men AND preferred a 50-50?_____

   d.   How many parents surveyed were men OR preferred a 50-50?_____

NAME _____ CLASS _____ DATE _____

Reteaching
4.5 Multiplication Principle of Counting

◆ **Skill A** Drawing and using tree diagrams

Recall Each branch of a tree diagram represents a possible outcome. The number of branches at each endpoint depends on the number of ways that an event can happen. Together, all of the branches represent all of the possible outcomes of an event.

◆ **Example 1**
Three flavors of ice cream, four toppings, cherries, and chocolate sprinkles are available for making sundaes. Use a tree diagram to determine how many different sundaes can be made with one flavor of ice cream, one topping, and either cherries or chocolate sprinkles.

◆ **Solution**
Start with one point and make three branches to represent the three flavors of ice cream.

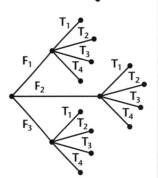

From each point at the end of these three branches, make four branches to represent the four toppings.

From each point at the end of the 12 branches, make two branches to represent the cherries and chocolate sprinkles.

Count the number of paths in the tree. There are 24 different sundaes that can be made with the given ingredients.

An alternative is to count the points at the ends of the paths.

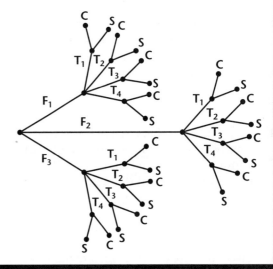

Use a tree diagram to answer each question.

1. When packing for a business trip, Mr. Simmons selects 3 suits, 4 shirts, and 5 ties, all of which can be used interchangeably. How many different outfits can he put together? _____

2. The science class decided to present a pet display for the kindergartners. Two students volunteered to bring fish, 3 could bring cats, 3 wanted to bring dogs, and 2 said they could bring rabbits. If the display could include one pet from each category, how many displays were possible? _____

◆ **Skill B** Using the Multiplication Principle of Counting

Recall The total number of ways to choose A AND B AND C = number of ways to choose A · number of ways to choose B · number of ways to choose C

◆ **Example 1**
When Marianne got a new job, she bought 5 tops, 4 skirts, and 3 pairs of shoes, all of which could be used interchangeably for work. How many days could Marianne go to work without wearing the same outfit?

◆ **Solution**
To put an outfit together, there are 5 tops to choose from. Once a top is chosen, there are 4 skirts to choose from. Once a top and skirt are chosen, there are 3 pairs of shoes to choose from.

Number of outfits =
number of ways to choose a top · number of ways to choose a skirt · number of ways to choose a pair of shoes $5 \cdot 4 \cdot 3 = 60$ outfits

---

Use the Multiplication Principle of Counting to solve each of the following:

3. If the new license plate pattern in the 51st state is any letter, followed by any digit, followed by any letter, how many distinct license plates can be made? _____

4. The house numbers in a new development are to be three-digit numbers that will range from 700 to 899 with the additional stipulation that the last digit must be even. How many possible house numbers will there be? _____

5. Two cards are drawn from a regular deck of 52 playing cards one at a time, without replacement. How many ways are there to draw two kings? _____

Reteaching
4.6 Theoretical Probability

◆ **Skill A** Finding the probability that an event will occur

Recall The theoretical probability, P, that an event will occur is defined as $P = \frac{s}{n}$, where s is the number of successful outcomes and n is the total number of equally likely outcomes.

> ◆ **Example 1**
> Find the probability that in one roll of a 12-sided number cube (dodecahedron) a prime number will result. Express your answer as a fraction, a decimal, and a percent.
>
> ◆ **Solution**
> Each of the 12 faces is equally likely to land face up on any one roll. Since 2, 3, 5, 7, and 11 are the prime numbers between 1 and 12, there are 5 ways for a successful outcome to occur.
>
> $$P = \frac{\text{number of successes}}{\text{total number of outcomes}} = \frac{s}{n} = \frac{5}{12}$$
>
> $$\frac{5}{12} = 0.41\overline{6} \approx 41.6\%$$

Find each probability as fraction, a decimal, and a percent.

1. From a bag containing 12 marbles, one is drawn randomly. If the bag contains 2 yellow marbles, 3 green marbles, 4 blue marbles, and 3 red marbles, find the probability of drawing a green marble. _____

2. If a letter is selected at random from the word *probability,* find the probability that the letter is a consonant. _____

3. If a letter is selected at random from all the letters in this question, find the probability that the letter selected is a *t*. _____

◆**Skill B**   Using counting principles to find the probability that an even will occur

Recall   Addition Counting Principle: The number of ways to choose A OR B = number of ways to choose A + number of ways to choose B − the number of ways to choose A AND B.

Multiplication Counting Principle: The number of ways to choose A AND B = number of ways to choose A · number of ways to choose B

◆ **Example**

The table presents the results of a class survey concerning a possible dance.

	Favor dance	Oppose dance	Total
Boys	53	42	95
Girls	74	28	102
Total	127	70	197

A student who participated in the survey is chosen at random. What is the probability that the student

a. is a girl?

b. opposes the dance?

c. is a girl AND opposes the dance?

d. is a girl OR opposes the dance?

◆ **Solution**

$$P = \frac{\text{number of successes}}{\text{total number of outcomes}}$$

The denominator for each event is the total number of students who participated in the survey, 197.

a. Since a total of 102 girls participated in the survey, the probability that the student chosen is a girl is $\frac{102}{197}$.

b. Since the total number of students opposing a dance is 70, the probability of choosing a student who opposes a dance is $\frac{70}{197}$.

c. The total number of students who are both girls AND oppose having a dance is found where the Girls row and the Oppose dance column Intersect, or 28. The probability of choosing a student who is both a girl and opposes the dance is $\frac{28}{197}$.

d. By the Addition Counting Principle, the number of students who are girls or oppose the dance = number of students who are girls + number of students opposing a dance − number who are girls and oppose a dance
$102 + 70 − 28 = 144$
The probability of choosing a student who is either a girl or opposes the dance is $\frac{144}{197}$.

A survey concerning school uniforms was conducted among parents. The results are shown in the table.

A parent who participated in the survey is selected at random. What is the probability that the parent

	For uniforms	Against uniforms	Total
Fathers	43	24	67
Mothers	29	49	78
Total	72	73	145

4. is a mother? _____

5. is a mother AND is against school uniforms? _____

6. is against school uniforms? _____

7. is a mother OR is against school uniforms? _____

Reteaching
4.7 Independent Events

◆ **Skill A**   Defining independent events

Recall   In independent events, the occurrence of the first event does not affect the probability that the second event will occur.

◆ **Example**

Determine whether the following are independent or dependent:

a. A number cube is rolled, and a coin is tossed. What is the probability that the number cube displays an odd number and the coin lands tails up?

b. Two cards are drawn from a standard deck without replacement (that is, the first card is not put back before the second is drawn). What is the probability that both cards are kings?

◆ **Solution**

a. Tossing a coin has no relationship to rolling a number cube. Thus, the probability of the coin landing tails up is not affected by the outcome of the number cube roll. The events are independent.

b. Of the 52 cards in a standard deck, 4 are kings, so the probability of drawing a king is $\frac{4}{52}$, or $\frac{1}{13}$. If a king is drawn first and not replaced, there are 3 kings remaining out of 51 cards, so on the second draw, the probability of drawing a king is $\frac{3}{51}$, or $\frac{1}{17}$. Thus, the probability of the second event (drawing a king) depends on the occurrence of the first event (drawing a king). The events are dependent.

Determine whether each of the following are independent or dependent events:

1. Two marbles are drawn from a bag without replacement. If the bag contains 4 green marbles, 5 red marbles, and 3 blue marbles, what is the probability of drawing 2 red marbles?   _____

2. A card is drawn from a standard 52-card deck and a number cube is rolled. What is the probability of drawing a queen and rolling a 5?   _____

3. Two number cubes are rolled. What is the probability that the sum of the numbers rolled is 9 and that the second cube shows a 3?   _____

4. A card is drawn from a standard 52-card deck and replaced. A second card is then drawn. Find the probability that two aces are drawn.   _____

◆ **Skill B**   Finding the probability of independent events

Recall   If events *A* and *B* are independent, then the probability of *A* AND *B* = probability of *A* · probability of *B*.

◆ **Example 1**

Andre and Angela play darts as a team against other two-person teams. If Andre makes a bull's-eye 70% of the time and Angela makes a bull's-eye 60% of the time, find the probability that both will hit a bull's-eye when making one throw apiece.

◆ **Solution**

The probability that Andre makes a bull's-eye on any shot is 70% = $\frac{70}{100}$ = $\frac{7}{10}$.

The probability that Angela makes a bull's-eye on any shot is 60% = $\frac{60}{100}$ = $\frac{6}{10}$.

Since the events are independent, the probability that both Andre and Angela make a bull's-eye is $\frac{7}{10} \cdot \frac{6}{10} = \frac{42}{100} =$ 42%.

This can be pictured on a 10 × 10 square grid, where 70 of the squares represent Andre's probability of making a bull's-eye and 60 of the squares represent Angela's probability. These regions overlap on 42 squares, which represents the probability that both make a bull's-eye.

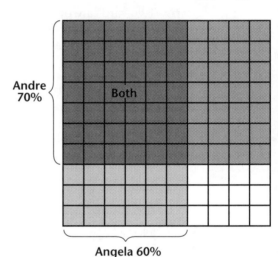

Find the probability of each situation described.

5. Two number cubes are rolled. Find the probability that both cubes show a 3 or less.   _____

6. Two cards are drawn from a standard 52-card deck. The first is replaced before the second is drawn. Find the probability that both cards drawn are spades.   _____

7. A coin is flipped and a number cube is rolled. Find the probability that the coin shows heads and the cube shows a 6.   _____

8. The digits from 0 to 9 are printed on 10 Ping-Pong balls and placed in a bag. Two balls are drawn, with the first ball being replaced before the second is drawn. What is the probability of drawing the digits 7 and 5, in that order, on the two draws?   _____

Reteaching
5.1 Functions and Relations

◆ Skill A Understanding differences between relations and functions

Recall A relation is a set of ordered pairs.
A function is a relation in which each x-coordinate is paired with one and only one y-coordinate.

◆ Example
State whether each set of ordered pairs is a function.
a. $\{(3, 1), (4, 6), (0, 1), (3, 2), (-4, 6), (5, 2)\}$
b. $\{(1,0), (-5,9), (4, 6), (2, 3), (5, 8), (-2, 4), (0, 4)\}$

◆ Solution
a. If all x-coordinates are different, the set names a function. If any x-coordinates repeat, check their y-coordinates. Since $(3, 1)$ and $(3, 2)$ have the same x-coordinate but different y-coordinates, the relation is not a function.
b. Because all the x-coordinates are different, the set names a function.

Decide whether each set represents a function.

1. $\{(3, 5), (2, 6), (4, 6), (-1, 6), (5, 8)\}$ _____

2. $\{(1, 0), (1, 4), (1, -3)\}$ _____

3. $\{(-6, 8), (6, 8), (-5, 7), (5, 7)\}$ _____

4. $\{(5, 1), (4, 2), (3, 3), (2, 4)\}$ _____

◆ Skill B Using the $f(x)$ function notation to represent and evaluate functions

Recall
$$f(x) = \underset{\uparrow}{a}x^3 + \underbrace{bx^2 + cx + \ldots}$$

replacement variable   function rule
Substitute the value of the replacement variable into the function rule and simplify.

◆ Example
Let $f(x) = 5x^3 - 3x^2 + 2x$. Find $f(3)$.

◆ Solution
Replace each x in the function rule with 3.
$f(3) = 5(3)^3 - 3(3)^2 + 2(3)$
$\quad\quad = 5(27) - 3(9) + 2(3)$ Exponents first
$\quad\quad = 135 - 27 + 6$ Multiplication before addition and subtraction
$\quad\quad = 108 + 6$ Subtraction and addition from left to right
$f(3) = 114$

Evaluate each function.

5. Let $f(x) = 7x^2 - 5x$. Find $f(-2)$. _____

6. Let $g(x) = -x^2 + 2x$. Find $g(4)$. _____

Evaluate each function when x is -3.

7. $f(x) = -2x^3 - x$ _____

8. $g(x) = \dfrac{1}{x^2}$ _____

◆ **Skill C** Testing graphs to identify functions by using the vertical-line test

  Recall Two points whose x-coordinates are the same, but whose y-coordinates are different, lie on the same vertical line.

◆ **Example**
  Use the vertical-line test to decide whether each graph represents a function.

  a.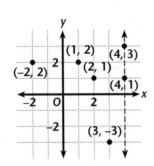

  b.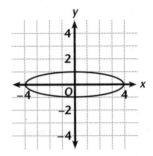

◆ **Solution**
  a. The relation shown has six points. Two of these points lie on the vertical line $x = 4$. Therefore, the graph does not represent a function.
  b. Because any vertical line drawn between $x = -4$ and $x = 4$ will cross the graph at more than one point, the graph does not represent a function.

Use the vertical-line test to decide whether each graph represents a function.

9.

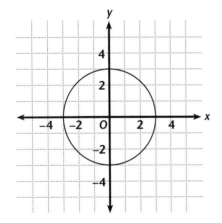

10.

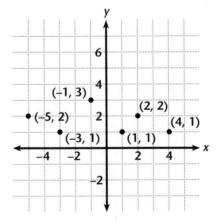

 # Reteaching
5.2 Exploring Transformations

◆ **Skill A** Identifying basic transformations of parent functions

Recall A parent function is the most basic of a family of functions. When a parent function is stretched, reflected, or shifted, it is said to be transformed.

◆ **Example**

The absolute-value parent function $y = |x|$ and the table of values used to graph it are shown.

x	-4	-3	-2	-1	0	1	2	3	4
y	4	3	2	1	0	1	2	3	4

Draw the graph of each related function listed below. Then describe how the graph of the parent function has been transformed.

a. $y = 2|x|$ **b.** $y = |x + 1|$ **c.** $y = -|x| + 2$

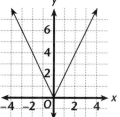

◆ **Solution**

a. Use the table of values for the parent function to make a table for $y = 2|x|$.
Multiply each value in the second row by 2.

x	-4	-3	-2	-1	0	1	2	3	4
y	8	6	4	2	0	2	4	6	8

The graph has been stretched by a factor of 2.

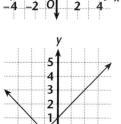

b. Make a new table of values for the function $y = |x + 1|$.

x	-4	-3	-2	-1	0	1	2	3	4
y	3	2	1	0	1	2	3	4	5

The graph has been shifted 1 unit to the left.

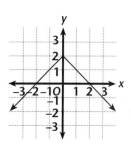

c. Use the table values for the parent function to make a table for $y = -|x| + 2$.
Change each value in the second row to its opposite and add 2.

x	-4	-3	-2	-1	0	1	2	3	4
y	-2	-1	0	1	2	1	0	-1	-2

The graph has been reflected through the x-axis and shifted 2 units up.

Identify each parent function and name the transformation(s) applied to it. Then draw and label the graph of each transformed function.

1. $y = -|x|$ _____

2. $y = 3|x|$ _____

3. $y = |x - 3|$ _____

4. $y = |x| - 1$ _____

5. $y = -2|x + 1|$ _____

6. $y = -3|x| - 3$ _____

◆ **Skill B** Identifying stretches, reflections, and shifts as they apply to the graphs of functions

Recall Multiplying a function by -1 reflects it through the x-axis.
Multiplying a function by a positive number stretches it.
Multiplying a function by a negative number stretches it and reflects it through the x-axis.
Adding or subtracting a number to a function shifts a function vertically.
Adding or subtracting a number to the independent variable, x, shifts the function horizontally.

◆ **Example**
Identify the parent function of $y = 3x^2$ and tell what transformation is applied to it. Draw and label the graph of the parent function and of the transformed function.

◆ **Solution**
The parent function is $y = x^2$. The coefficient 3 indicates that the function $y = x^2$ is stretched vertically by 3.
First draw the graph of the parent function.
Make a table of values and graph.

x	-4	-3	-2	-1	0	1	2	3	4
y	16	9	4	1	0	1	4	9	16

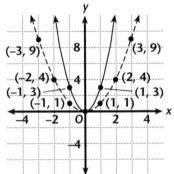

Next multiply each value in the second row by 3 to make a table of values for the stretched function. Plot the stretched function on the same coordinate plane as the parent function.

x	-4	-3	-2	-1	0	1	2	3	4
y	48	27	12	3	0	3	12	27	48

Identify each parent function and name the transformation applied to it. Then draw and label the graph of the parent function and the graph of the transformed function on the grid provided.

7. $y = |x| - 3$ **8.** $y = \left(\frac{1}{2}\right)x^2$ **9.** $y = 2|x - 3|$ **10.** $y = (x + 3)^2$

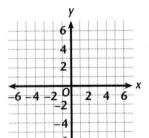
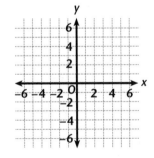
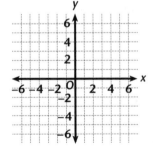
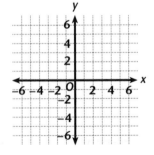

 **Reteaching**
5.3 Stretches

◆**Skill A** Describing the effects of stretches on the graphs of functions

Recall Multiplying a function by a scale factor greater than 1 produces function values that are larger than those of the parent function; multiplying by a scale factor between 0 and 1 produces function values that are smaller than those of the parent function.

◆ **Example 1**
Identify the parent function and the scale factor for $y = 6|x|$. Describe the effect of the stretch and sketch the graph transformed function.

◆ **Solution**
The parent function is the absolute-value function, $y = |x|$. The scale factor is 6, the coefficient of the $|x|$ term. Because the scale factor is greater than 1, the y-values of the stretched function will be 6 times larger than the y-values of the parent function. The graph of the function will rise or fall more steeply as it is stretched vertically. First sketch the graph of the parent function. Recall its V-shape and remember that its two branches extend through the first and second quadrants.

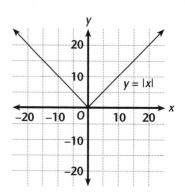

Next choose a point on the parent function. Then apply the scale factor to determine a point on the stretched graph. For example, (2, 2) becomes (2, 12). Sketch the resulting graph.

◆ **Example 2**
Describe the effect of the stretch for the function $y = \frac{|x|}{5}$.

◆ **Solution**
The parent function is $y = |x|$ and the scale factor is

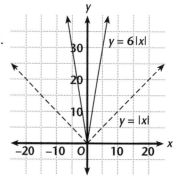

$\frac{1}{5}$. The y-values of the stretched function will be $\frac{1}{5}$ of the y-values of the parent function. The V-shape of the graph will rise or fall less steeply than that of the parent function.

Identify each parent function and scale factor.

1. $y = \frac{3}{x}$ _____

2. $y = \frac{x^2}{5}$ _____

3. $y = 3|x|$ _____

4. $y = 2(10^x)$ _____

Describe the effect of the stretch on each function and sketch the resulting graph on graph paper.

5. $y = 3 \cdot 2^x$ _____

6. $y = \frac{x^2}{4}$ _____

◆ **Skill B** Identifying coefficients to determine the amount of stretch

Recall Division can be expressed as multiplication by the reciprocal: $\frac{x}{a} = \frac{1}{a} \cdot x$.

◆ **Example**

For each function, identify the coefficient. Then determine the amount and direction of stretch produced by the scale factor.

a. $y = \frac{1}{4x}$ **b.** $y = \frac{4}{x}$

◆ **Solution**

a. The function can be rewritten as $y = \frac{1}{4} \cdot \frac{1}{x}$.

The coefficient is $\frac{1}{4}$. Every y-value of the parent function is multiplied by $\frac{1}{4}$, so each point of the parent function $y = \frac{1}{x}$ is moved to $\frac{1}{4}$ of the distance from the x-axis.

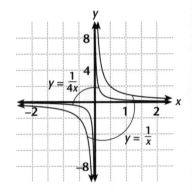

b. The function can be rewritten as $y = 4 \cdot \frac{1}{x}$, so the coefficient is 4.

Each y-value of the parent function is multiplied by 4. The positive y-values, which lie in the first quadrant, are 4 times their original values. The negative y-values, which are in the third quadrant, are also 4 times their original values. In the resulting graph, each point is 4 times the distance from the x-axis.

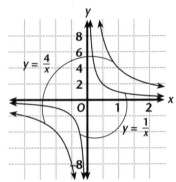

Given each function, identify the coefficient. Then determine the amount of the stretch produced by the scale factor.

7. $y = 2x$ _____

8. $y = 3|x|$ _____

9. $y = \frac{5}{x}$ _____

10. $y = \frac{|x|}{3}$ _____

11. $y = \frac{x}{4}$ _____

12. $y = \frac{x^2}{5}$ _____

13. $y = \frac{2|x|}{5}$ _____

14. $y = 0.65|x|$ _____

Reteaching
5.4 Reflections

◆ **Skill A**   Understanding the connections between reflections of a graph
and the opposite or negative of a function

Recall   The points (a, b) and $(a, -b)$ are equidistant from the x-axis
along a line perpendicular to the x-axis. These points are
vertical reflections of each other through the x-axis.

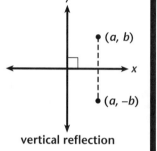

vertical reflection

◆ **Example 1**
Determine whether $y = -x^2$ is a vertical reflection of
the parent function $y = x^2$.

◆ **Solution**
In a vertical reflection, every point (a, b) of the function
is matched by its reflection point $(a, -b)$.
Compare the table of values for the two functions.

x	-5	-4	-3	-2	-1	0	1	2	3	4	5
y	25	16	9	4	1	0	1	4	9	16	25

x	-5	-4	-3	-2	-1	0	1	2	3	4	5
y	-25	-16	-9	-4	-1	0	-1	-4	-9	-16	-25

Since $-x^2 = (-1) \cdot x^2$, each y-value of the second function is the opposite of
each y-value of the first function. Every point (a, b) on $y = x^2$ is matched by
its reflection point $(a, -b)$ on $y = -x^2$, so $y = -x^2$ is a vertical reflection of the
parent function.

◆ **Example 2**
Determine whether $y = |x| - 1$ is a vertical reflection of the parent function
$y = |x|$.

◆ **Solution**
If the function $y = |x| - 1$ is the vertical reflection of $y = |x|$, each point (a, b)
of the first function will be matched by its reflection point $(a, -b)$. Choose an
x-value to test. It is best not to choose 0 or 1 as a test value.
For $x = 3$:

$$y = |x| - 1 \qquad y = |x|$$
$$y = |3| - 1 \qquad y = |3|$$
$$y = 3 - 1 \qquad y = 3$$
$$y = 2$$

Since $(3, 2)$ is not a vertical reflection of $(3, 3)$ through the x-axis, the function
$y = |x| - 1$ is not a vertical reflection of $y = |x|$.

Determine whether each function is a vertical reflection of its parent function.

1. $y = -|x|$ _____

2. $y = 3x - 2$ _____

3. $y = 2^{-x}$ _____

4. $y = x^2 - 4$ _____

5. $y = -10^x$ _____

6. $y = \dfrac{-1}{x}$ _____

◆ **Skill B**   Identifying minimum or maximum values of absolute-value and quadratic functions

Recall   The minimum value of a function is its smallest value; the maximum value of a function is its largest value. For a quadratic function and an absolute-value function, the turning point, or vertex, of the graph is the maximum or minimum value of the function.

◆ **Example 1**
Determine the location of the minimum or maximum value of the function $y = -2|x|$.

◆ **Solution**
The parent function is $y = |x|$. Its graph opens upward and has a minimum value. The coefficient of 2 stretches the graph vertically by a scale factor of 2. The negative sign causes the graph to be reflected through the x-axis, so its graphs opens downward and therefore has a maximum value. To find the maximum value, make a table of values.

x	−6	−5	−4	−3	−2	−1	0	1	2	3	4	5	6
$y = -2\|x\|$	−12	−10	−8	−6	−4	−2	0	−2	−4	−6	−8	−10	−12

The table shows that the greatest y-value occurs at (0, 0). Therefore, the maximum value of $y = -2|x|$ is 0.

◆ **Example 2**
Find the maximum or minimum value of the function $y = x^2 - 3$.

◆ **Solution**
The parent function is $y = x^2$. Its graph opens upward and has a minimum value. Subtraction of 3 from each y-value shifts the graph downward 3 units, so the minimum is 3 units below the minimum of the parent function. Since $y = x^2$ has a minimum value of 0, $y = x^2 - 3$ has a minimum value of −3.

Identify each function as having a maximum or minimum value. Determine the location of that point for each function.

7. $y = -5|x|$ _____

8. $y = x^2 - 2$ _____

9. $y = -(x^2 + 2)$ _____

10. $y = 3|x|$ _____

11. $y = |x| - 3$ _____

12. $y = -x^2 - 1$ _____

 **Reteaching**

5.5 Translations

◆ **Skill A** Describing the effects of translations on the graphs of functions

Recall When a function is translated vertically or horizontally, every point is moved the same distance in the same direction.

◆ **Example 1**
A point on the graph of $y = |x| + 3$ is $(-4, 7)$. What is the resulting point if the function is translated vertically by -4?

◆ **Solution**
Since every point is moved the same distance in the same direction, the point is moved down 4 units. A vertical change affects only the y-value of a point, so the new y-value is $7 - 4$, or 3. Therefore, the resulting point is $(-4, 3)$.

◆ **Example 2**
Choose a point on the graph of $y = x^2 - 2$. Then find the corresponding point after the function is translated horizontally by 3.

◆ **Solution**
Choose any value to substitute for x in the equation. If $x = 4$, for example, $y = 4^2 - 2$, or $y = 14$. The coordinates of the point are $(4, 14)$. Because a horizontal translation affects only the x-coordinate of a point, 3 must be added to the first coordinate of the point $(4, 14)$. The new value of x is $4 + 3$, or 7. The coordinates of the point corresponding to $(4, 14)$ after the horizontal translation are $(7, 14)$.

Using the function $y = |x| + 4$, find each point after the indicated translation.

1. $(-2, 6)$, vertical by -2 _____

2. $(5, 9)$, horizontal by 4 _____

3. $(-7, 11)$, horizontal by -3 _____

4. $(3, 7)$, vertical by -5 _____

5. $(13, 17)$, vertical by 6 _____

6. $(-5, 9)$, horizontal by -12 _____

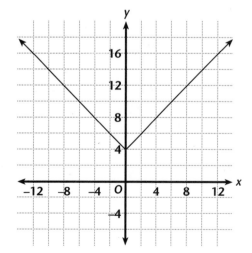

◆ **Skill B** Identifying relationships between the translation of a graph and the addition or subtraction of a constant

Recall The general formula for the translation of the parent function $y = f(x)$ is $y = a(x - h) + k$. For this to work you must *subtract h* from the independent variable and *add k* to the function. A positive h-value shifts the function to the right, and a negative h-value shifts the function to its left. Likewise, a positive k-value shifts the function upward, and a negative k-value shifts the function downward. A positive a-value stretches the function by a factor of a, and a negative a-value stretches the function by a factor of a and reflects it through the x-axis. For example, $y = -3(x + 3)^2 + 5$, or $y = -3(x - (-3))^2 + 5$, shifts the function $y = x^2$ 3 units to the left and 5 units up, stretches the function by a factor 3, and reflects it through the x-axis.

◆ **Example**

Identify the parent function. Describe the effect of the addition or subtraction on the parent function.

a. $y = x^2 + 4$ **b.** $y = |x + 5|$

◆ **Solution**

a. The parent function is $y = x^2$. Once each x-value is squared, the result is then increased by 4. The effect of this addition on the parent function is a shift 4 units up.

b. The parent function is the absolute-value function, $y = |x|$. Before taking the absolute value of each x, the x-value is increased by 5 or decreased by -5. The effect of this subtraction on the parent function is a shift 5 units to the left.

Identify each parent function and describe the effect of the addition or subtraction on the parent function.

7. $y = 2^x + 3$ _____

8. $y = \dfrac{1}{x} - 2$ _____

9. $y = x^2 + 2$ _____

10. $y = |x| - 4$ _____

11. $y = 10^{(x-2)}$ _____

12. $y = |x + 3|$ _____

13. $y = (x - 4)^2$ _____

14. $y = x^2 - 3$ _____

Reteaching
5.6 Combining Transformations

◆ **Skill A**   Identifying parent functions in transformations

Recall   Transformations of a function are indicated by the addition or subtraction of constants from the variable term or from the entire function or by multiplication or division of the variable term by a constant.

◆ **Example**
In the following equations, identify the parent function.
  a. $y = -3|x - 2| + 5$ **b.** $y = -(x + 3)^2 - 4$

◆ **Solution**
  a. Identify the additions, multiplications, subtractions, or divisions that occur. If the addition of 5 is removed, the equation becomes $y = -3|x - 2|$. If the multiplication by -3 is removed, the equation becomes $y = |x - 2|$. Finally, if the subtraction of 2 is removed, the equation becomes $y = |x|$. This is the absolute-value parent function.

  b. Start with $y = -(x + 3)^2 - 4$, and remove the additions and subtractions, starting with the subtraction of 4 outside the parentheses. This leaves $y = -(x + 3)^2$. Then remove the negative sign preceding the parentheses, leaving $y = (x + 3)^2$. Finally, remove the addition of 3 within the parentheses, producing the function $y = x^2$. This is the quadratic parent function.

Identify the parent function for each.

1. $y = -2|x + 1| - 4$ _____

2. $y = 3(x - 1)^2 - 2$ _____

3. $y = 3 \cdot 2^{-x} + 1$ _____

4. $y = -3(x + 2) - 4$ _____

5. $y = \dfrac{3}{x} + 2$ _____

6. $y = \dfrac{3}{x + 2}$ _____

7. $y = 3x^2 - 4$ _____

8. $y = -2(x - 1)^2$ _____

◆ **Skill B**   Understanding the effect of order on combining transformations

Recall   To determine the order of transformation to a function, reverse the order of operations. Addition or subtraction indicates a vertical translation; multiplication or division indicates a vertical stretch; addition or subtraction within parentheses or absolute-value symbols indicates a horizontal translation.

◆ **Example**
Describe the various transformation included in the equation $y = 2|x - 1| + 3$.

◆ **Solution**
The first operation to consider is the addition of 3. This affects the parent function by translating it vertically 3 units up. The second operation, multiplication by 2, stretches the translated function by a factor of 2. The third operation, subtraction of 1, translates the stretched function horizontally to the right 1 unit. Thus, the parent function, $y = |x|$, has been shifted to the right 1 unit, stretched by a factor of 2, and then shifted 3 units up.

Describe the transformations of the parent functions included in each equation.

9. $y = -3|x + 2| - 3$ _____

_____

_____

10. $y = 2(x - 3)^2 + 1$ _____

_____

_____

11. $y = 4|x - 1| + 2$ _____

_____

_____

12. $y = 4 \cdot 2^x - 2$ _____

_____

_____

Reteaching
6.1 Exploring Exponents

◆ **Skill A** Understanding the concepts of exponents and powers

Recall In the expression $2^4 = 16$, the 2, known as the base, is multiplied by itself 4 times. Thus, $2 \cdot 2 \cdot 2 \cdot 2 = 2^4 = 16$.

◆ **Example 1**
Express 10^7 in customary notation.

◆ **Solution**
The base, 10, is multiplied by itself 7 times. Each time a number is multiplied by 10, the product contains another zero. Therefore, the expression 10^7 is a 1 followed by 7 zeros, or 10,000,000.

◆ **Example 2**
Express 28,431 in exponential notation.

◆ **Solution**
It is essential to know the value of each decimal place when writing a number in expanded form.
$$28{,}431 = 2 \cdot 10{,}000 + 8 \cdot 1000 + 4 \cdot 100 + 3 \cdot 10 + 1$$
In exponential form, each power of ten is written with an exponent to indicate the number of zeros in the power.
$$2 \cdot 10^4 + 8 \cdot 10^3 + 4 \cdot 10^2 + 3 \cdot 10^1 + 1$$

Write each exponential expression in customary notation.

1. 10^6 _____

2. 10^9 _____

3. 5^4 _____

4. 3^5 _____

5. 4^3 _____

6. 2^8 _____

Write each expansion in exponential notation.

7. $3 \cdot 3 \cdot 3 \cdot 3$ _____

8. $4 \cdot 4 \cdot 4 \cdot 4 \cdot 4 \cdot 4$ _____

9. $6 \cdot 6 \cdot 6 \cdot 6 \cdot 6 \cdot 6 \cdot 6$ _____

10. $10 \cdot 10 \cdot 10 \cdot 10$ _____

11. $5 \cdot 5 \cdot 5$ _____

12. $2 \cdot 2 \cdot 2 \cdot 2 \cdot 2 \cdot 2$ _____

Use exponents to express each number in exponential notation.

13. 43,795 _____

14. 145,286 _____

15. 723,621 _____

16. 36,291 _____

◆ **Skill B** Using the properties of exponents to simplify expressions

Recall If x is any number and a is an integer greater than 1, then

$$x^a = \underbrace{x \cdot x \cdot x \cdot x \cdot \ldots \cdot x}_{a \text{ factors}}.$$

When $a = 1$, $x^a = x^1 = x$.

◆ **Example 1**

Simplify each expression.

a. $5^3 \cdot 5^4$ **b.** $\dfrac{4^5}{4^3}$

◆ **Solution**

a. Rewrite each power as a product of the base factors as indicated by the exponent.

$$5^3 = 5 \cdot 5 \cdot 5 \qquad 5^4 = 5 \cdot 5 \cdot 5 \cdot 5$$

Then, $5^3 \cdot 5^4 = 5 \cdot 5 \cdot 5 \cdot 5 \cdot 5 \cdot 5 \cdot 5$

The product can be expressed by using a base of 5 and an exponent of 7. That is, $5^3 \cdot 5^4 = 5^7$ or $5^3 \cdot 5^4 = 5^{3+4} = 5^7$.

b. If the numerator is expressed as the product of base factors it becomes $4^5 = 4 \cdot 4 \cdot 4 \cdot 4 \cdot 4$. The denominator becomes $4^3 = 4 \cdot 4 \cdot 4$. Then $\dfrac{4 \cdot 4 \cdot 4 \cdot 4 \cdot 4}{4 \cdot 4 \cdot 4}$ can be written as $\dfrac{4}{4} \cdot \dfrac{4}{4} \cdot \dfrac{4}{4} \cdot 4 \cdot 4$, or

$1 \cdot 1 \cdot 1 \cdot 4 \cdot 4 = 4^2$. The quotient $\dfrac{4^5}{4^3}$ is 4^2, or $\dfrac{4^5}{4^3} = 4^{5-3} = 4^2$.

◆ **Example 2**

Simplify the expression $(5^2)^3$.

◆ **Solution**

The expression within the parentheses, 5^2, is being used as a factor 3 times, so $(5^2)^3 = 5^2 \cdot 5^2 \cdot 5^2$. Each factor is the second power of 5 and can be written as $5 \cdot 5$. The expression $(5^2)^3$ is $5 \cdot 5 \cdot 5 \cdot 5 \cdot 5 \cdot 5$, or $5^{2 \cdot 3} = 5^6$.

Simplify each expression.

17. $4^5 \cdot 4^3$ _____

18. $3^6 \cdot 3^5$ _____

19. $\dfrac{3^{10}}{3^2}$ _____

20. $\dfrac{7^8}{7^5}$ _____

21. $(4^2)^4$ _____

22. $(5^3)^2$ _____

23. $(3^5)^3$ _____

24. $(9^4)^3$ _____

25. $2^5 \cdot 2^3 \cdot 2^7$ _____

26. $\dfrac{10^5}{10^3}$ _____

27. $(7^5 \cdot 7^2)^3$ _____

28. $\dfrac{5^5 \cdot 5^3}{5^4}$ _____

29. $(3^{10})^5$ _____

30. $6^3 \cdot 6^2 \cdot 6^5$ _____

 **Reteaching**

6.2 Multiplying and Dividing Monomials

◆**Skill A** Understanding the structure and concept of monomials

Recall When a number and letter are written together with no operation sign between them, multiplication is indicated.
A *monomial* is either a constant, a variable, or a product of a constant and one or more variables.

◆ **Example**
Identify which expressions are monomials.

    a. $-5m^2n^3$ **b.** $4x^2 - 5y^2$ **c.** 2

◆ **Solution**
    a. Since the expression can be written completely as the product of factors, it is a monomial.

$$-5m^2n^3 = -5 \cdot m \cdot m \cdot n \cdot n \cdot n$$

    b. The expression cannot be written only as a product of factors. Two products are being subtracted. It is not a monomial.

$$4x^2 - 5y^2 = 4 \cdot x \cdot x - 5 \cdot y \cdot y \qquad \text{subtraction}$$

    c. The constant, 2, can be considered a product because it can be written as $2 \cdot 1$. Therefore, it is a monomial.

Determine whether each expression is a monomial. Write *yes* or *no*.

1. $2 - 5t^2$ _____

2. $3x^2y^3$ _____

3. -6 _____

4. $-15xyz^3$ _____

5. $3x^2 - 5x + 2$ _____

6. $\frac{3a^2}{b^2}$ _____

7. $4 + 2$ _____

8. $5c^3$ _____

9. t _____

10. $(5a^2)^3$ _____

11. $-4x^6y^2z^5$ _____

12. $(3a^2b^2)^3 2(ab)^2$ _____

13. $5x^2 + 6x - 7$ _____

14. $\frac{2x^2 - 7}{3}$ _____

15. $5^3 - 5s$ _____

16. $-6t^3(2t^3)^2$ _____

17. $3y^2 + 2y - 1$ _____

18. $(-5x^2y^2z)^5$ _____

19. $(3 - t^2)^2$ _____

20. $5a^5$ _____

◆ **Skill B** Using the properties of exponents to simplify expressions

Recall Multiplication is indicated by a variety of symbols. In arithmetic, an "×" is used. In algebra, a raised dot (·), parentheses, or no symbol at all can indicate multiplication. The Associative Property of Multiplication allows you to organize factors so that similar factors can be multiplied; that is, $(a \cdot b) \cdot c = a \cdot (b \cdot c)$. The Commutative Property of Multiplication allows you to change the order of the factors without changing the product; that is, $a \cdot b = b \cdot a$.

◆ **Example**
Simplify each expression.
   a. $6(5x^2y^3)^4$ **b.** $(rs^2)^3(r^2s^4)^2$

◆ **Solution**
   a. Only the product within the parentheses is affected by the exponent 4. The factor 6 is used only once. If the expression were expanded, it would be written $6 \cdot 5x^2y^3 \cdot 5x^2y^3 \cdot 5x^2y^3 \cdot 5x^2y^3$. Then, using the Commutative Property of Multiplication to reorder the factors, it is possible to put numerical factors together and like variable factors together.

$$6 \cdot 5 \cdot 5 \cdot 5 \cdot 5 \cdot x^2 \cdot x^2 \cdot x^2 \cdot x^2 \cdot y^3 \cdot y^3 \cdot y^3 \cdot y^3$$

   Multiply the numerical factors to get 3750 and add the exponents of like bases to get x^8y^{12}. In simplified form, the expression is $3750x^8y^{12}$.

   b. One expression is used three times as a factor and the other is used twice. The expression becomes $rs^2 \cdot rs^2 \cdot rs^2 \cdot r^2s^4 \cdot r^2s^4$. If the bases are reordered by using the Commutative Property, the expression becomes $r \cdot r \cdot r \cdot r^2 \cdot r^2 \cdot s^4 \cdot s^4 \cdot s^4 \cdot s^4$. If a factor has no exponent, the exponent is assumed to be 1. Then the exponents of the like bases are added, resulting in r^7s^{14}.

Simplify each expression.

21. $5(2x^3y^2)^4$ _____

22. $(-4m^3n)^5$ _____

23. $(a^2b)^3(ab^3)^2$ _____

24. $-2(5ab^2)^3(2a^2b)^4$ _____

25. $-(3mn^2)^3(-2mn)^2$ _____

26. $(3a^2b)^2(5ab^2)^3(4a^2b^2)^4$ _____

27. $-3(r^2s^3t)^2(-2rs^2t^3)^2rs$ _____

28. $x^2y^2(-2x^2y)^3(-2xyz^2)^2$ _____

29. $(-4r^2s^3t)^3(-2r^2s)^2(-2t)^3$ _____

30. $(5m^2n^2)^3(-2m^3n)^2(-3m^2)^3$ _____

31. $(-2x^2y^3)^2(-3x^5y^3)(-2x^2)^2$ _____

32. $(3a^5b^2)^2(-2a^2b)^3(-2a^4b^2)^2$ _____

33. $(8r^5t^2)^2(-2s^2t)^3$ _____

34. $(4m^2n^5)^2(-m^2n)^3(-m)^5$ _____

35. $(-4a^2b^5)^2(-a^2b^3)^2(-a^5b)^3$ _____

36. $(-2x^5y^2)^2(-3x^2y^3)^3$ _____

NAME _____ CLASS _____ DATE _____

 **Reteaching**
6.3 Negative and Zero Exponents

◆ **Skill A** Understanding the concepts of negative and zero exponents

Recall The relationship between an exponent and a power can be seen as a pattern.
Examining the powers of 2 and their corresponding exponents reveals a pattern:

2^4	2^3	2^2	2^1
16	8	4	2

As the exponent decreases by 1, the value of the power decreases by a factor of $\frac{1}{2}$.

◆ **Example 1**
Evaluate 3^0.

◆ **Solution**
Examining powers of 3 with positive exponents, it is apparent that as the exponent decreases by 1, the power decreases by a factor of $\frac{1}{3}$.

3^4	3^3	3^2	3^1
81	27	9	3

Notice that as the exponent decreases by 1, the value of the power has been divided by 3 or multiplied by a factor of $\frac{1}{3}$.

If 3^0 is added to the list, it becomes obvious that $3^0 = 1$.

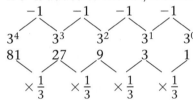

Therefore, $3^0 = 1$.

Because this same pattern can be formed for any positive or negative base, x, it can be shown that $x^0 = 1$.

◆ **Example 2**
Evaluate 3^{-2}.

◆ **Solution**
Continuing the pattern of Example 1, it is obvious that as the exponent decreases by 1, the power decreases by a factor of $\frac{1}{3}$.

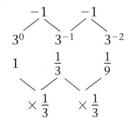

Then, $3^{-2} = \frac{1}{9}$.
Comparing 3^{-2} with 3^2, it is apparent that when exponents are opposites, resulting powers are reciprocals. If $3^3 = 27$, then $3^{-3} = \frac{1}{27}$.

Evaluate each expression.

1. 4^0 _____   **2.** 5^{-2} _____   **3.** 8^0 _____   **4.** 4^{-1} _____

5. 3^{-3} _____   **6.** 1^{-2} _____   **7.** 5^{-3} _____   **8.** 4^{-3} _____

◆ **Skill B**   Simplifying expressions containing negative and zero exponents

    Recall   To add two integers with the same sign, add their absolute vales and keep their common sign. To add two integers with different signs, subtract their absolute values and use the sign of the number with the greater absolute value. Subtraction is the same as adding the opposite.

    ◆ **Example 1**
    Simplify the expression $y^{-5} \cdot y^3$.

    ◆ **Solution**
    Multiplication of powers of the same base is done by adding exponents. The bases are both y, so the exponents -5 and 3 must be added. Therefore, $y^{-5} \cdot y^3 = y^{-5+3} = y^{-2}$.

    ◆ **Example 2**
    Simplify the expression $\dfrac{m^4}{m^7}$.

    ◆ **Solution**
    Division of powers of the same base is done by subtracting exponents. Therefore, $\dfrac{m^4}{m^7} = m^{4-7} = m^{-3}$.

    ◆ **Example 3**
    Simplify $c^{-3} \cdot c^0$.

    ◆ **Solution**
    This expression represents the product of the powers of the same base. Therefore, the product is found by adding the exponents. Then, $c^{-3} \cdot c^0 = c^{-3+0} = c^{-3}$. Alternatively, $c^0 = 1$, because any base to the zero power equals 1. A factor multiplied by 1 is itself. Thus, $c^{-3} \cdot c^0 = c^{-3} \cdot 1 = c^{-3}$.

Simplify each expression.

9. $a^3 \cdot a^{-5}$ _____

10. $c^2 \cdot c^{-7}$ _____

11. $\dfrac{y^3}{y^6}$ _____

12. $\dfrac{m^{-3}}{m^6}$ _____

13. $p^8 \cdot p^0$ _____

14. $q^0 \cdot q^{-5}$ _____

15. $x^{-8} \cdot x^{-3}$ _____

16. $z^{-5} \cdot z^8$ _____

17. $\dfrac{t^{-5}}{t^{-10}}$ _____

18. $5^{-3} \cdot 5^8$ _____

19. $x^5 \cdot x^{-3} \cdot x^{-7}$ _____

20. $3^3 \cdot 3^{-10} \cdot 3^6$ _____

21. $\dfrac{t^{-5} \cdot t^5}{t^3}$ _____

22. $\dfrac{4^7}{4^{-3}}$ _____

23. $5^3 \cdot 5^0 \cdot 5^{-1}$ _____

24. $a^2 \cdot a^{-5}$ _____

25. $\dfrac{r^{10} \cdot r^{-2}}{r^5}$ _____

26. $\dfrac{2^{10} \cdot 2^{-10}}{2^{10}}$ _____

Reteaching
6.4 Scientific Notation

◆ **Skill A**   Expressing numbers in scientific notation

Recall   To multiply a number by a power of 10, move the decimal point to the right.
To divide a number by a power of 10, move the decimal point to the left.

◆ **Example**
Express the following numbers in scientific notation:

a. 2,300,000,000   **b.** 0.0000052

◆ **Solution**
a. Because scientific notation requires a number between 1 and 10, the decimal
point is placed between the 2 and 3, giving 2.3. Count how many places
the decimal point was moved to form 2.3 in order to find the exponent of
10. The decimal point was moved 9 places to the left. In scientific notation,
$2,300,000,000 = 2.3 \times 10^9$.

b. To form a number between 1 and 10, the decimal point is moved between
the 5 and 2, giving 5.2. The decimal point was moved 6 places to the right.
Thus, in scientific notation, $0.0000052 = 5.2 \times 10^{-6}$.

**Express each of the following in scientific notation. Explain
what you did.**

1. 4,570,000,000 _____

2. 0.0000023 _____

3. 0.00458 _____

4. 62,000,000 _____

5. 70,500,000,000 _____

6. 0.0000875 _____

7. 5800 _____

8. 0.026 _____

9. 35,000,000 _____

10. .000000072 _____

11. 2,070,000,000,000 _____

12. .00305 _____

◆ **Skill B**   Performing computations with scientific notation

Recall   A power of 10 written in exponential form indicates the number of zeros in the number when written in customary form. That is, 10^4 means 4 zeros in the number 10,000.

◆ **Example 1**
Perform the following computations:
  a. $(7 \times 10^3)(5 \times 10^4)$ **b.** $\dfrac{7 \times 10^7}{2 \times 10^3}$

◆ **Solution**
  a. To multiply numbers in scientific notation, first multiply the whole or decimal factors. In this example, $7 \times 5 = 35$. Then multiply the powers of 10 by adding exponents. $10^3 \cdot 10^4 = 10^7$. The product 35×10^7 should be changed to scientific notation since 35 is not between 1 and 10.
$$35 \times 10^7 = 3.5 \times 10^1 \times 10^7 = 3.5 \times 10^8$$

  b. To divide numbers in scientific notation, first divide the whole or decimal factors. In this example, $7 \div 2 = 3.5$. Then divide the powers of 10 by subtracting exponents. $10^7 \div 10^3 = 10^4$. The quotient is 3.5×10^4. This result is in scientific notation because the decimal factor is between 1 and 10.

◆ **Example 2**
Perform the following computations with a calculator:
  a. (2.34 E 02)(1.7 E 03)   **b.** $\dfrac{5.6 \text{ E } 05}{2.8 \text{ E } 02}$

◆ **Solution**
  a. When numbers are written in scientific notation on a calculator, the power of 10 is expressed in E notation, where E is followed by the exponent of 10. 2.34 E 02 means 2.34×10^2. This number can be entered on the calculator by the following succession of keys: 2.34 [EE] 2. Enter the multiplication sign and the second number in the same form. The result is 3.978 E 05, or 3.978×10^5.

  b. Enter the numerator in E notation. Use the [EE] key for E. Enter the division sign and follow it with the denominator. The quotient is 2 E 03, or 2×10^3.

Perform each computation without a calculator. Express each answer in scientific notation.

13. $(8 \times 10^4)(3 \times 10^3)$ _____   **14.** $\dfrac{6.5 \times 10^5}{1.3 \times 10^3}$ _____

15. $(4.3 \times 10^2)(5.2 \times 10^4)$ _____   **16.** $\dfrac{8.4 \times 10^6}{2.1 \times 10^2}$ _____

Perform each computation with a calculator.

17. (5.63 E 03)(4.2 E 04) _____   **18.** $\dfrac{7.5 \text{ E } 04}{2.5 \text{ E } 02}$ _____

19. (4.25 E 05)(6.32 E 06) _____   **20.** $\dfrac{1.2 \text{ E } 05}{6 \text{ E } 02}$ _____

Reteaching
6.5 Exponential Functions

◆ **Skill A** Understanding exponential functions

Recall When a whole-number base is used with a positive integral exponent, the power is greater than or equal to the base. When a whole-number base is used with a nonpositive integral exponent, the power is less than the base.

◆ **Example**
For each equation,
- describe the effect on y as x increases from zero;
- describe the y- value at $x = 0$;
- describe the effect on y as x decreases from zero; and
- graph the equation.

  a. $y = 3^x$ **b.** $y = 0.3^x$

◆ **Solution**

a. Because the base 3 is greater than 1, the effects of x can be seen in a chart.

x	-3	-2	-1	0	1	2	3	4
$y = 3^x$	$\frac{1}{27}$	$\frac{1}{9}$	$\frac{1}{3}$	1	3	9	27	81

From the chart, it is apparent that as x increases from zero, the function value gets larger. At $x = 0$, y is equal to 1, and as x decreases from zero, the values of the function gets smaller, approaching zero.

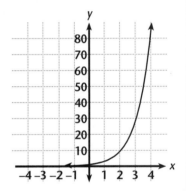

b. Because the base is less than 1, the effects of x will be different from the previous example.

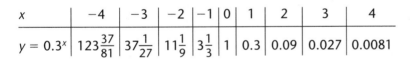

x	-4	-3	-2	-1	0	1	2	3	4
$y = 0.3^x$	$123\frac{37}{81}$	$37\frac{1}{27}$	$11\frac{1}{9}$	$3\frac{1}{3}$	1	0.3	0.09	0.027	0.0081

From this chart, it is obvious that as x increases from zero, the function values decrease. At $x = 0$, the y value is still equal to 1, and as x decreases from zero, the values of the function get larger.

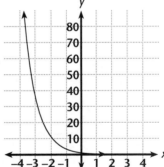

For each function, describe the effect on y as x increases from negative to positive. Sketch the graph on graph paper.

1. $y = 5^x$ _____

2. $y = 0.5^x$ _____

3. $y = 8^x$ _____

4. $y = 0.8^x$ _____

◆ **Skill B**   Identifying graphs of exponential functions for different bases

Recall   The graph of an exponential function with a base greater than 1 increases as x increases; the graph of an exponential function whose base is greater than 0 but less than 1 decreases as x increases.

◆ **Example**
Determine the function represented by each graph.

a.
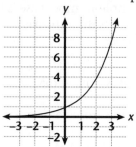

b.
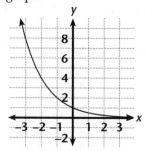

◆ **Solution**

a. Because the value of y increases as x increases, the base can be assumed to be greater than 1. The point with coordinates (0, 1) is on the graph, so the function is of the form $y = b^x$. To determine b, check the value of y for $x = 1$. Since $y = 2$ when $x = 1$, the base is 2 and the function is $y = 2^x$.

b. Because the value of y decreases as x increases, the base must be a number between 0 and 1. The point with coordinates (0, 1) is on the graph, so this function is also of the form $y = b^x$. When $x = 1$, $y = \frac{1}{2}$, so the base is $\frac{1}{2}$ and the function is $y = \left(\frac{1}{2}\right)^x$.

Determine the function for each graph.

5.
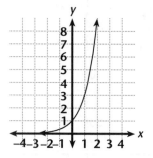

_____

6.
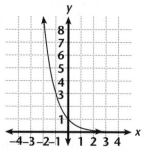

_____

7.
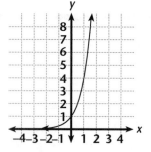

_____

8.
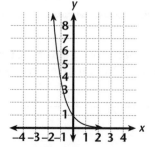

_____

Reteaching
6.6 Applications of Exponential Functions

◆ **Skill A** Using exponential functions to model applications

Recall An exponential function results when a value increases by a constant multiplier.

◆ **Example**
Model the following problem as an exponential function:
A stock has been increasing in value by 5% each year since its purchase 4 years ago. If it was purchased at $62, what is its present value?

◆ **Solution**
If the initial value of the stock was $62, it increased in value by 5% of $62 after the first year, or $62 + 62 \cdot 0.05$. By the Distributive Property, this expression can be rewritten as $62(1 + 0.5)$. The factor $(1 + 0.05)$ is the multiplier, which increases the new value by 0.05, or 5%. Therefore, the value of the stock after 4 years can be written as follows:

$$62 \quad \cdot \quad (1 + 0.05) \quad \cdot \quad (1 + 0.05) \quad \cdot \quad (1 + 0.05) \quad \cdot \quad (1 + 0.05)$$

↑	↑	↑	↑	↑
initial value	value after year 1	value after year 2	value after year 3	value after year 4

More succinctly, this product can be written as $62 \cdot (1 + 0.05)^4$ which equals approximately 75.36.

As an exponential function, the expression could be written as $V = I(1 + r)^t$, where V means value, I is initial investment, r stands for annual rate of increase, and t is the number of years of investment.

Model each problem as an exponential function.

1. A baseball card that originally cost $24 has been decreasing in value by 6% annually for the last 5 years. What is its present value?

_____

2. A large city has a population of 500,000 and has been increasing in size by 2% annually for 6 years. What was its population 6 years ago?

_____

3. One share of stock that originally sold for $54 has increased in value by 4% annually for the last 7 years. What is its present value?

_____

◆ **Skill B**   Using exponential functions to solve problems about growth and decay

Recall   When an expression contains more than one operation, including parentheses and exponents, the expression within parentheses is simplified first.

◆ **Example**
If $2000 is deposited in a savings account paying 7% interest compounded quarterly, what will the account be worth in 8 years if no deposits or withdrawals are made? (Use the compound interest formula $A = P\left(1 + \frac{r}{n}\right)^{nt}$, where A is the amount after t years, P is the principal invested, r is the annual interest rate in decimal form, and n is the number of times the interest is compounded per year.)

◆ **Solution**
Substituting the values supplied into the formula, the equation becomes $A = 2000\left(1 + \frac{0.07}{4}\right)^{4 \times 8}$. Recognize that compounding quarterly means that interest is applied 4 times per year, so $n = 4$ in this example. Simplifying $\frac{0.07}{4}$, it becomes 0.0175. Then, $A = 2000(1 + 0.0175)^{4 \times 8}$, or $A = 2000(1.0175)^{32}$. Next, the exponential factor is evaluated. This power is approximately 1.7422. Multiplying this result and rounding to the nearest cent, $A = 3484.40$. In 8 years the account will be worth $3484.40.

Solve each problem.

4. If $5000 is invested in an account paying 6% interest compounded semiannually, how much will the investment be worth in 10 years? (Use $A = P\left(1 + \frac{r}{n}\right)^{nt}$.)

_____

5. A bacterial culture contains N bacteria after t hours according to the formula $N = S(2.54)^{0.04t}$, where S is the starting number of bacteria. How many bacteria will be present after 24 hours if the initial number is 12,000?

_____

6. The population of a certain town decreases according to the formula $P = 8000(0.9)^{0.2t}$, where t represents the number of years after 1990. What will the population be in the year 2000?

_____

Reteaching
7.1 Exploring Polynomial Functions

◆ **Skill A** Evaluating polynomial expressions

Recall The expression $x^2 + 5x + 6$ is an example of a polynomial. A polynomial is an expression that consists of one or more monomials. To evaluate $x^2 + 5x + 6$ for any number, replace x with the given value each time it appears in the expression and simplify.

◆ **Example 1**
Evaluate $x^2 + 5x + 6$ when x is 2.

◆ **Solution**
Substitute 2 for x.

$$x^2 + 5x + 6 = (2)^2 + 5(2) + 6$$
$$= 4 + 10 + 6$$
$$= 20$$
When x is 2, $x^2 + 5x + 6$ is 20.

◆ **Example 2**
Evaluate $a^2 - 2a + 4$ when a is -3.

◆ **Solution**
$$a^2 - 2a + 4 = (-3)^2 - 2(-3) + 4$$
$$= 9 + 6 + 4$$
$$= 19$$
When a is -3, $a^2 - 2a + 4$ is 19.

Evaluate each polynomial for the given value of the variable.

1. $3a^2 + 5$ when a is -2 _____

2. $5c^2 - 2c$ when c is 3 _____

3. $-m^2 + 2$ when m is -7 _____

4. $-2p^2 - 10$ when p is 4 _____

5. $t^2 - 2t + 4$ when t is 2 _____

6. $3s^2 + 3s - 5$ when s is -2 _____

7. $-5b^2 + 5b + 5$ when b is 0 _____

8. $4c^2 + 6c + 2$ when c is -5 _____

9. $2t + 3t^2 + 3$ when t is -1 _____

10. $9 - 5q + q^2$ when q is 3 _____

11. $6 - d^2 - 3d$ when d is -6 _____

12. $-4 - 3x^2 - 5x$ when x is 4 _____

13. $3y^2 - 6y + 2$ when y is $\frac{1}{3}$ _____

14. $\frac{2}{3}a^2 + 2a + 4$ when a is $\frac{1}{2}$ _____

15. $-2z^2 - z + 3$ when z is $-\frac{1}{2}$ _____

16. $\frac{1}{2} - \frac{1}{3}n - n^2$ when n is $\frac{5}{6}$ _____

◆ **Skill B** Evaluating polynomial functions

Recall An equation written in the form $f(x) = x^2 - x - 2$ is a polynomial function. To find $f(x)$ for any value of x, substitute the value for x each time it appears in the function and simplify.

◆ **Example**
Evaluate the function $f(x) = x^2 - x - 2$ when x is 2.

◆ **Solution**
Substitute 2 for x.
$f(2) = (2)^2 - 2 - 2$
$\quad = 4 - 2 - 2$
$\quad = 0$
When x is 2, $f(x) = x^2 - x - 2$ is 0.

Evaluate each function when x is 3.

17. $f(x) = 2x^2 + 3x - 9$ _____

18. $h(x) = -3x^2 - 2x + 9$ _____

For each function, find $g(-2)$.

19. $g(x) = 5x^2 - 8x + 10$ _____

20. $g(x) = 3 - 5x - 2x^2$ _____

◆ **Skill C** Evaluating volume and surface area functions

Recall The volume of a box with a square base of length x and height 10 can be represented by the function $V(x) = 10x^2$.
The surface area of the same box can be represented by the function $S(x) = 2x^2 + 40x$.

◆ **Example**
Find the surface area of a box with a square base if the side length of the square is 3 inches and the height of the box is 10 inches.

◆ **Solution**
Substitute 3 for x in the surface area function. In other words, find $S(3)$.
$S(x) = 2x^2 + 40x$
$S(3) = 2(3)^2 + 40(3)$
$\quad = 18 + 120$
$\quad = 138$
The surface area of the box is 138 square inches.

A box has a square base of 5 meters on each side. If the height is 3 meters, the volume and surface area of the box are determined by $V(x) = 3x^2$ and $S(x) = 2x^2 + 12x$. Find the volume and surface area.

21. volume _____

22. surface area _____

A box has a square base of 2.5 feet on each side. If the height is 20 feet, the volume and surface area of the box are determined by $V(x) = 20x^2$ and $S(x) = 4x^2 + 180x$. Find the volume and surface area.

23. volume _____

24. surface area _____

Reteaching
7.2 Adding and Subtracting Polynomials

◆ **Skill A**   Adding polynomials

Recall   To add two polynomials, add the coefficients of the like terms.

◆ **Example 1**
Add the polynomials horizontally.
$3a^3 + 2a^2 + a + 5$ and $2a^3 + 4a - 6$

◆ **Solution**
Group like terms.
$(3a^3 + 2a^2 + a + 5) + (2a^3 + 4a - 6)$
$= (3a^3 + 2a^3) + 2a^2 + (a + 4a) + (5 - 6)$
$= 5a^3 + 2a^2 + 5a - 1$
The sum of $3a^3 + 2a^2 + a + 5$ and $2a^3 + 4a - 6$ is $5a^3 + 2a^2 + 5a - 1$.

◆ **Example 2**
Add the same two polynomials vertically.

◆ **Solution**
Line up the variables. Use zero for the coefficient of any missing variable.
$$3a^3 + 2a^2 + 1a + 5$$
$$\underline{+\ 2a^3 + 0a^2 + 4a - 6}$$
$$5a^3 + 2a^2 + 5a - 1$$
The sum of $3a^3 + 2a^2 + a + 5$ and $2a^3 + 4a - 6$ is $5a^3 + 2a^2 + 5a - 1$.

Add each pair of polynomials.

1. $5b^2 + 3b$ and $b^2 - 2b$ _____

2. $8c^2 - 2c$ and $2c^2 + 3c$ _____

3. $b^3 + 2b^2 + 3b$ and $4b^3 - 5b^2 + 4b$ _____

4. $3y^3 + 3y - 1$ and $2y^3 + 5y^2 + 3y$ _____

5. $5r^2 + 3r + 6$ and $2r^3 + r^2 + 4r$ _____

6. $4m^3 - 5m^2 - m$ and $3m^3 - 3m - 5$ _____

7. $(2x^2 + 3x + 4) + (-5x^2 + x - 7)$ _____

8. $(x^2 - x + 6) + (3x^2 - x + 3)$ _____

9. $(2x^2 + 3x + 6) + (-2x^2 - 7)$ _____

10. $(4x^3 - 5x + 4) + (3x^3 + 5x - 3)$ _____

◆ **Skill B**   Finding the opposites of polynomials

   Recall   To find the opposite of a term, change the sign in front of the term.

   ◆ **Example**
   Find the opposite of $2b^2 + 3b - 7$.

   ◆ **Solution**
   The opposite of $2b^2 + 3b - 7$ is $-(2b^2 + 3b - 7)$.
   $-(2b^2) = 2b^2$; $-(3b) = -3b$; $-(-7) = 7$
   Thus, $-(2b^2 + 3b - 7) = -2b^2 - 3b + 7$.

Find the opposite of each polynomial.

11. $3c^2 + c + 5$ _____

12. $n^2 - 2n + 3$ _____

13. $-2z^2 - z - 1$ _____

14. $5r^2 + 4r - 9$ _____

15. $4t^2 - t$ _____

16. $-9q^2 - q - 3$ _____

17. $5 - 2a - 3a^2$ _____

18. $5e^3 - 4e^2 + 2e$ _____

◆ **Skill C**   Subtracting polynomials

   Recall   To subtract a polynomial, add its opposite.

   ◆ **Example**
   Subtract $2c^2 - 3c - 5$ from $5c^2 - 2c + 3$.

   ◆ **Solution**
   $(5c^2 - 2c + 3) - (2c^2 - 3c - 5) = (5c^2 - 2c + 3) + (-2c^2 + 3c + 5)$
   $= 3c^2 + c + 8$
   $2c^2 - 3c - 5$ subtracted from $5c^2 - 2c + 3$ is $3c^2 + c + 8$.

Subtract each pair of polynomials.

19. $(x^2 + 3x + 2) - (3x^2 + x - 6)$ _____

20. $(2x^2 - 5x + 1) - (2x^2 + 3x - 2)$ _____

21. $(3x^2 + x - 4) - (-3x^2 + 2x - 7)$ _____

22. $(-x^2 - x - 4) - (-2x^2 - 4x + 3)$ _____

Perform each indicated operation.

23. $(2x^2 + 3x - 4) - (2x - 5) + (x^2 - x + 1)$ _____

24. $(-5x^2 - 2x + 1) - (3x^2 + 4x - 2) - (-8x^2 - 5x - 3)$ _____

Reteaching
7.3 Exploring Multiplication Models

◆**Skill A** Using models to multiply monomials and binomials

Recall Algebra tiles can be used to find the product of 3 and $x + 1$. Three positive 1-tiles are placed along the left line. One positive x-tile and one positive 1-tile are placed along the top line. Place tiles that form a rectangular shape inside the lines.

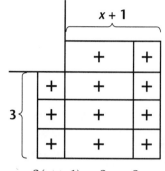

$$3(x + 1) = 3x + 3$$

◆ **Example 1**
Use algebra tiles to find the product $2(x + 2)$.

◆ **Solution**

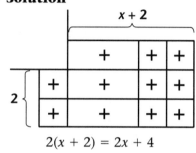

$$2(x + 2) = 2x + 4$$

◆ **Example 2**
Use algebra tiles to find the product $x(x - 1)$.

◆ **Solution**

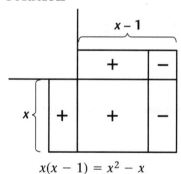

$$x(x - 1) = x^2 - x$$

Use algebra tiles to help you find each product.

1. $2(x + 3) =$ _____

2. $3(2x - 2) =$ _____

3. $x(x - 3) =$ _____

4. $x(-x + 1) =$ _____

5. $x(2x + 4) =$ _____

6. $x(3x - 3) =$ _____

7. $2x(x - 2) =$ _____

8. $2x(3x + 2) =$ _____

◆**Skill B**   Using models to multiply two binomials

Recall   Algebra tiles can also be used to find the product of two binomials.

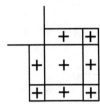

This model shows that $(x + 1)(x + 1) = x^2 + 2x + 1$.

◆ **Example 1**
Use algebra tiles to find the product $(x - 2)(x - 3)$.

◆ **Solution**

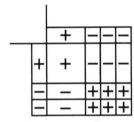

$(x - 2)(x - 3) = x^2 - 5x + 6$

◆ **Example 2**
Use algebra tiles to find the product $(x + 2)(x - 2)$.

◆ **Solution**

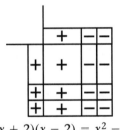

$(x + 2)(x - 2) = x^2 - 4$

Use algebra tiles to help find each product.

9. $(x + 2)(x + 1)$ _____

10. $(x + 2)(x - 1)$ _____

11. $(x - 2)(x + 1)$ _____

12. $(x - 2)(x - 1)$ _____

13. $(x + 3)(x + 3)$ _____

14. $(x - 3)(x - 3)$ _____

15. $(x + 3)(x - 3)$ _____

16. $(x - 3)(x + 3)$ _____

17. $(x + 1)(2x + 3)$ _____

18. $(x - 2)(2x - 1)$ _____

19. $(3x + 2)(x - 2)$ _____

20. $(2x + 1)(3x - 2)$ _____

Reteaching
7.4 Exploring Multiplication of Binomials

◆ **Skill A** Multiplying monomials and binomials

Recall The Distributive Property can be used to find the product of a monomial and a binomial.

    ◆ **Example**
    Use the Distributive Property to find the product $x(x - 4)$.

    ◆ **Solution**
    $x(x - 4) = x(x) - x(4)$
             $= x^2 - 4x$
    The product $x(x - 4)$ is $x^2 - 4x$.

Use the Distributive Property to find each product.

1. $4(x + 5)$ _____

2. $5(x - 2)$ _____

3. $x(2x - 2)$ _____

4. $2x(3x + 1)$ _____

5. $-5x(x - 6)$ _____

6. $-3x(-x - 3)$ _____

◆ **Skill B** Multiplying two binomials

Recall The Distributive Property can also be used to multiply two binomials.

    ◆ **Example**
    Use the Distributive Property to find the product $(x + 2)(x + 5)$.

    ◆ **Solution**
    $(x + 2)(x - 5) = (x + 2)(x) - (x + 2)(5)$
                   $= x(x) + (2)(x) - (x)(5) - (2)(5)$
                   $= x^2 + 2x - 5x - 10$
                   $= x^2 - 3x - 10$

Use the Distributive Property to find each product.

7. $(x + 1)(x + 4)$

8. $(x + 3)(x + 2)$

9. $(x + 5)(x - 3)$

_____ _____ _____

10. $(2x + 3)(x + 2)$

11. $(x - 5)(3x - 3)$

12. $(3x - 4)(4x - 3)$

_____ _____ _____

◆ **Skill C** Using the FOIL method to multiply two binomials

Recall To multiply two binomials:
multiply the **F**irst terms;
multiply the **O**utside terms;
multiply the **I**nside terms;
add the outside and inside products; and
multiply the **L**ast terms.

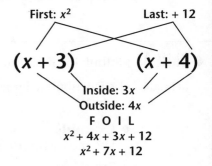

First: x^2 Last: $+ 12$

$(x + 3) \quad (x + 4)$

Inside: $3x$
Outside: $4x$
F O I L
$x^2 + 4x + 3x + 12$
$x^2 + 7x + 12$

◆ **Example 1**
Use the FOIL method to find the product
$(x + 3)(x - 4)$.

◆ **Solution**

$$
\begin{array}{ccccc}
 & \textbf{F} & \textbf{O} & \textbf{I} & \textbf{L} \\
(x + 3)(x - 4) = & (x)(x) & - (x)(4) & + (3)(x) & - (3)(4) \\
= & x^2 & - 4x & + 3x & - 12 \\
= & x^2 - x - 12 & & &
\end{array}
$$

The product $(x + 3)(x - 4)$ is $x^2 - x - 12$.

◆ **Example 2**
Use the FOIL method to find the product $(2x - 3)(3x - 1)$.

◆ **Solution**

$$
\begin{array}{cccc}
 & \textbf{F} & \textbf{O} \;\; \textbf{I} & \textbf{L} \\
(2x - 3)(3x - 1) = & 6x^2 & - 2x - 9x & + 3 \\
= & 6x^2 - 11x + 3 & &
\end{array}
$$

The product $(2x - 3)(3x - 1)$ is $6x^2 - 11x + 3$.

Find each product. Use the space provided.

13. $(x + 2)(x + 5)$

14. $(x + 3)(x - 4)$

15. $(x - 5)(x - 3)$

_____

_____

_____

16. $(2x - 3)(x + 3)$

17. $(2x - 5)(2x - 5)$

18. $(5x - 4)(5x - 4)$

_____

_____

_____

19. $(4x + 1)(x - 1)$

20. $(2x - 3)(x - 2)$

21. $(-3x + 4)(2x - 3)$

_____

_____

_____

Reteaching
7.5 Common Factors

◆**Skill A** Listing prime numbers

Recall A positive integer is a prime number if it has exactly two factors, the numbers 1 and itself. The number 3 is prime because it has exactly two factors: 1 and 3. The number 4 is not prime because it has three factors: 1, 2, and 4.

◆ **Example**
Which of the following whole numbers are prime?
a. 1   b. 37   c. 87

◆ **Solution**
a. 1 is not prime because it has only one factor, itself.
b. 37 is prime because it has only two factors: 1 and 37.
c. 87 is not prime because it has more than two factors: 1, 3, 29, and 87.

List the factors of each whole number. Tell whether each number is prime.

1. 12 _____   2. 35 _____   3. 47 _____

4. 57 _____   5. 77 _____   6. 97 _____

◆**Skill B** Factoring out a common monomials

Recall The product of x and $x + 3$ is $x^2 + 3x$. To factor an expression such as $x^2 + 3x$ means to factor out the greatest common factor (GCF) of each term. The Distributive Property is used to rewrite the expression as the product $x(x + 3)$. Some expressions cannot be factored. Since the greatest common factor of $x^2 - 3$ is 1, $x^2 - 3$ is considered prime.

◆ **Example**
Factor each polynomial.
a. $4b^3 - 6b^2 + 10b$ b. $5y^3 - 15x^2 + 18xy$

◆ **Solution**
a. The GCF is $2b$.   b. The GCF is 1.
$4b^3 - 6b^2 + 10b = 2b(2b^2 - 3b + 5)$ $5y^3 - 15x^2 + 18xy$ is prime.

Factor each polynomial.

7. $3m^2 - 21m$ _____   8. $8t^2 + 15t$ _____

9. $18p^2 + 21p + 9$ _____   10. $4d^3 - 20d^2 + 8d$ _____

◆ **Skill C** Factoring common binomial factors

Recall When an expression has a common binomial factor, the Distributive Property is used to rewrite the expression as a product. Sometimes the expression must be regrouped first.

◆ **Example 1**
Factor $a(b - 1) + 3(b - 1)$.

◆ **Solution**
The common binomial factor is $b - 1$.
$a(b - 1) + 3(b - 1) = (b - 1)(a + 3)$

◆ **Example 2**
Factor $x^2 - 3x + 6x - 18$.

◆ **Solution**
Group the first two terms together. Group the last two terms together.
$(x^2 - 3x) + (6x - 18)$
Use the Distributive Property to factor the grouped terms.
$x(x - 3) + 6(x - 3)$
Factor out the common binomial.
$(x - 3)(x + 6)$
$x^2 - 3x + 6x - 18 = (x + 6)(x - 3)$

◆ **Example 3**
Factor $x^2 + 2x - 4x - 8$.

◆ **Solution**
$$x^2 + 2x - 4x - 8 = (x^2 + 2x) - (4x + 8)$$
$$= x(x + 2) - 4(x + 2)$$
$$= (x + 2)(x - 4)$$

Factor each polynomial.

11. $x(x + 5) + 8(x + 5)$

12. $(x - 7)x - (x - 7)3$

13. $x^2 + 3x + 4x + 12$

14. $x^2 - 3x - 2x + 6$

15. $x^2 + 5x - 2x - 10$

16. $x^2 + x - 4x - 4$

Reteaching
7.6 Factoring Special Polynomials

◆**Skill A** Factoring perfect-square binomials

Recall The binomial $(x - y)^2$ can be written as a trinomial by using the FOIL method.
$$(x - y)^2 = (x - y)(x - y)$$
$$= x^2 - xy - xy + y^2$$
$$= x^2 - 2xy + y^2$$

When a binomial is squared, the product has a pattern:
  The first term is one variable squared (x^2).
  The middle term is twice the product of the two variables ($2xy$).
  The second term is the second variable squared (y^2).
  If the binomial is a difference, the middle term is subtracted. If the binomial is a sum, the middle term is added.

The expression $x^2 + 2xy + y^2$ is called a perfect-square trinomial because it is written in the form $(x + y)^2$ when it is factored. If a trinomial has a perfect-square trinomial pattern, it can be factored by reversing the pattern.

◆ **Example**
Factor each expression.

  a. $s^2 + 6s + 9$ **b.** $4m^2 - 12mn + 9n^2$

◆ **Solution**
  a. The first term is a perfect square, s^2.
    The last term is a perfect square, $(3)^2$.
    The middle term, $6s$, is $(2)(s)(3)$.
    $s^2 + 6s + 9 = (s + 3)(s + 3)$
    $= (s + 3)^2$

  b. The first term is a perfect square, $(2m)^2$.
    The last term is a perfect square, $(3n)^2$.
    The middle term, $12mn$, is $(2)(2m)(3n)$.
    $4m^2 - 12mn + 9n^2 = (2m - 3n)(2m - 3n)$
    $= (2m - 3n)^2$

Square each expression.

1. $(x + 4)^2$ _____

2. $(2v - 5)^2$ _____

3. $(3d - a)^2$ _____

4. $(10k - t)^2$ _____

Factor each expression, if possible.

5. $x^2 + 2x + 1$ _____

6. $y^2 - 10y - 25$ _____

7. $4p^2 + 16p + 16$ _____

8. $4a^2 - 4ab + b^2$ _____

9. $r^2 - 4rs + 4s^2$ _____

10. $49b^2 - 42bc + 9c^2$ _____

◆ **Skill B**   Factoring the difference of two squares

Recall   The product $(x + y)(x - y)$ can be written as a binomial by using the FOIL method.
$(x + y)(x - y) = x^2 - xy + xy - y^2$
$= x^2 - y^2$
When the sum and difference of two monomials are multiplied, the product has a pattern:
        The first term is one variable squared (x^2).
        The second term is the other variable squared (y^2).
        The middle term is zero.
An expression written in the form $x^2 - y^2$ is called the difference of two squares. When a binomial is written as the difference of two squares, the binomial can be factored into the sum and difference of two monomials.

◆ **Example**
Factor each expression.

   a. $r^2 - 16$ **b.** $4b^2 - 9c^2$ **c.** $n^4 - 4$

◆ **Solution**
   a. The first term is a perfect square, $(r)^2$.
        The last term is a perfect square, $(4)^2$.
        $r^2 - 16 = (r)^2 - (4)^2$
                $= (r + 4)(r - 4)$

   b. The first term is a perfect square, $(2b)^2$.
        The last term is a perfect square, $(3c)^2$.
        $4b^2 - 9c^2 = (2b)^2 - (3c)^2$
                $= (2b + 3c)(2b - 3c)$

   c. The first term is a perfect square, $(n^2)^2$.
        The last term is a perfect square, $(2)^2$.
        $n^4 - 4 = (n^2)^2 - (2)^2$
                $= (n^2 + 2)(n^2 - 2)$

Find each product.

11. $(t + 5)(t - 5)$ _____   **12.** $(m - d)(m + d)$ _____

13. $(3r - s^2)(3r + s^2)$ _____   **14.** $(5p + 4q)(5p - 4q)$ _____

Factor each expression, if possible.

15. $t^2 - 49$ _____   **16.** $4 - a^2$ _____

17. $9d^2 + 1$ _____   **18.** $4s^2 - t^2$ _____

19. $25c^2 - 4q^2$ _____   **20.** $16b^2 - c^4$ _____

21. $m^2n^2 - p^2$ _____   **22.** $s^4 - t^4$ _____

Reteaching
7.7 Factoring Trinomials

◆ **Skill A** Factoring trinomials by using diagrams

Recall The product of $a + 2$ and $a - 5$ can be found by using the FOIL method.
$$(a + 2)(a - 5) = a^2 - 5a + 2a - 10$$
$$= a^2 - 3a - 10$$
To factor a trinomial such as $a^2 - 3a - 10$ means to write the expression as a product of monomials or binomials whose greatest common factor is one. If no monomials or binomials can be found, the trinomial is prime.

◆ **Example**
Use a diagram to factor $x^2 - 6x + 8$.

◆ **Solution**
First fill the boxes with what you know.

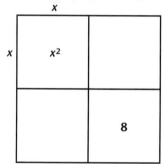

Complete the box with two factors that have a product of 8 and a sum of -6.

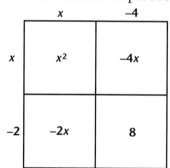

$$x^2 - 6x + 8 = (x - 2)(x - 4)$$

Use a diagram to factor each trinomial.

1. $x^2 + 3x + 2$ **2.** $x^2 + x - 12$ **3.** $x^2 - 10x + 21$

_____   _____   _____

4. $x^2 + 4x - 5$ **5.** $x^2 + 11x + 24$ **6.** $x^2 - 8x + 16$

_____   _____   _____

NAME _____ CLASS _____ DATE _____

◆ **Skill B**   Factoring trinomials by using lists

Recall   Another way to factor a trinomial such as $x^2 - 5x - 6$ is to first make a list of the pairs of factors of the constant. Then choose the right combination to complete the factors of the trinomial.

◆ **Example**
Use lists to factor $x^2 - 5x - 6$.

◆ **Solution**
List each pair of factors of -6 along with their sum.

Factors of -6	Sum of the factors
6 and -1	5
3 and -2	1
2 and -3	-1
1 and -6	-5

The sum of 1 and -6 is -5. Use the combination of 1 and -6 to form the factors.
$x^2 - 5x - 6 = (x + 1)(x - 6)$

Use lists to factor each trinomial.

7. $x^2 - x - 2$

8. $x^2 + 3x - 4$

9. $x^2 + 4x + 3$

_____

_____

_____

10. $x^2 - 4x + 3$

11. $x^2 + 2x - 8$

12. $x^2 + x - 20$

_____

_____

_____

13. $x^2 + 2x - 15$

14. $x^2 - 3x + 10$

15. $x^2 - x - 12$

_____

_____

_____

16. $x^2 + 6x + 8$

17. $x^2 - 20x + 36$

18. $x^2 + 2x - 24$

_____

_____

_____

NAME _____ CLASS _____ DATE _____

 Reteaching
8.1 Exploring Parabolas

◆ **Skill A** Finding the vertices and axes of symmetry of parabolas

Recall A quadratic function in the form $g(x) = a(x - h)^2 + k$, where $a \neq 0$, transforms the
parent function $g(x) = x^2$ by
- stretching the parent function by a factor of a,
- moving the vertex of the parent function from $(0, 0)$ to (h, k), and
- moving the axis of symmetry to $x = h$.

If $a > 0$, k is the minimum value of g.
If $a < 0$, k is the maximum value of g.

◆ **Example**
Determine the vertex, the axis of symmetry, and
the maximum or minimum, and sketch the graph
of $y = 3(x - 2)^2 - 4$.

◆ **Solution**
The value for h is 2.
The value for k is -4.
The value for a is 3.
The vertex is $(2, -4)$.
The axis of symmetry is $x = 2$.
The minimum value of y is -4.

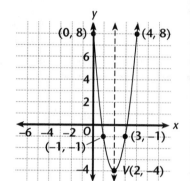

**Determine the vertex, the axis of symmetry, and the maximum
or minimum, and sketch the graph of each quadratic function.**

1. $y = 2(x - 5)^2 + 1$

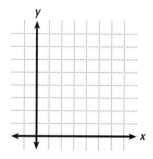

2. $y = -(x - 4)^2 - 2$

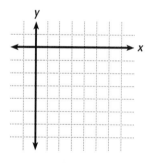

3. $y = -2(x + 1)^2 + 4$

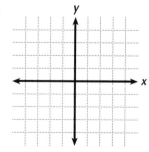

4. $y = (x - 3)^2 + 5$

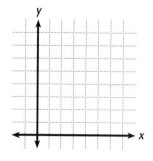

◆ **Skill B** Finding the zeros of quadratic functions

Recall The zeros of a quadratic function in the form $f(x) = ax^2 + bx + c$ are the x-values for which $f(x) = 0$. When a sketch of the function is drawn, the zeros of the function are the x-values of the points where the parabola intersects the x-axis.

◆ **Example**
Find the zeros of $y = x^2 - x - 6$ by graphing.

◆ **Solution**

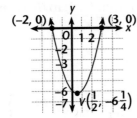

The zeros are 3 and -2.

Sketch the graph each quadratic function. Then find the zeros of each function.

5. $y = x^2 - 3x + 2$

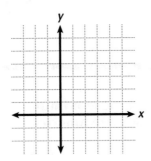

_____

6. $y = x^2 - 3x - 10$

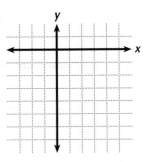

_____

7. $y = x^2 + 4x + 3$

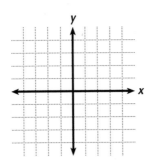

_____

8. $y = x^2 + 2x + 2$

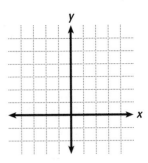

_____

9. $y = 2x^2 - 5x + 2$

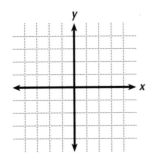

_____

10. $y = 2x^2 + x - 6$

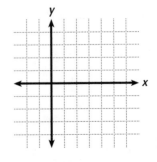

_____

 **Reteaching**

8.2 Solving Equations of the Form $x^2 = k$

◆ **Skill A** Finding square roots

Recall Every positive number has a positive and a negative square root. For example, the positive square root of 4 is 2 because $2 \cdot 2 = 4$. The negative square root of 4 is -2 because $-2 \cdot -2 = 4$. The positive root of 4 is indicated by $\sqrt{4}$. The negative square root of 4 is indicated by $-\sqrt{4}$.

Thus,

$$\sqrt{4} = 2 \text{ and } -\sqrt{4} = -2$$

◆ **Example**
Find each square root.

   a. $\sqrt{169}$ **b.** $-\sqrt{64}$ **c.** $\sqrt{45}$

◆ **Solution**

   a. $\sqrt{169} = 13$ **b.** $-\sqrt{64} = -8$ **c.** $\sqrt{45} \approx 6.71$

Find each square root. Round to the nearest hundredth when necessary.

1. $\sqrt{36}$ _____

2. $-\sqrt{196}$ _____

3. $-\sqrt{275}$ _____

4. $\sqrt{484}$ _____

5. $\sqrt{240}$ _____

6. $-\sqrt{841}$ _____

◆ **Skill B** Solving equations of the form $x^2 = k$, where $k \geq 0$

Recall There are two solutions to the equation $x^2 = 4$.

$$\sqrt{4} = 2 \text{ and } -\sqrt{4} = -2$$

If $x^2 = 4$, x is equal to 2 or -2; that is, $x = \pm 2$. The solutions are 2 and -2.

◆ **Example**
Solve each equation. Round to the nearest hundredth when necessary.

   a. $x^2 = 49$ **b.** $x^2 = 115$ **c.** $x^2 = 6.25$

◆ **Solution**

   a. $x = \pm 7$; the solutions are 7 and -7.
   b. $x = \pm 10.72$; the solutions are approximately 10.72 and -10.72.
   c. $x = \pm 2.5$; the solutions are 2.5 and -2.5.

Solve each equation. Round to the nearest hundredth when necessary.

7. $x^2 = 25$ _____

8. $x^2 = 75$ _____

9. $x^2 = 108$ _____

◆ **Skill C**   Solving equations of the form $(x + a)^2 = k$, where $k \geq 0$

Recall   An equation such as $(x + 3)^2 = 25$ is solved by using the following generalization:
If $x^2 = k$, and $k \geq 0$, then
- $x = \pm\sqrt{k}$, and
- the solutions are $\sqrt{k}$ and $-\sqrt{k}$.

When the generalization is applied to the equation $(x + 3)^2 = 25$, the result is
$x + 3 = \pm\sqrt{25} = \pm5$.
Thus, $x + 3 = 5$ and $x + 3 = -5$.
The solutions are 2 and -8.

◆ **Example**
Solve the equation $(x - 2)^2 - 9 = 0$.

◆ **Solution**
$$(x - 2)^2 - 9 = 0$$
$$(x - 2)^2 = 9$$
$$x - 2 = \pm3$$
$$x = 2 \pm 3$$
The solutions are $2 + 3$, or 5, and $2 - 3$, or -1.

Solve each equation in the space provided. Round to the nearest hundredth if necessary.

10. $(x + 1)^2 = 36$ _____

11. $(x - 4)^2 = 100$ _____

12. $(x + 3)^2 = 64$ _____

13. $(x + 7)^2 - 81 = 0$ _____

14. $(x - 2)^2 - 35 = 0$ _____

15. $(x - 1)^2 - 125 = 0$ _____

Reteaching
8.3 Completing the Square

◆ **Skill A** Completing the square

Recall The area of the shaded region is represented by $x^2 + 8x$. To complete the total area of the square, the area of the non-shaded region must be added to $x^2 + 8x$. The area to be added can be found by adding the square of $\frac{1}{2}$ of the coefficient of the x-term. Since the coefficient of x is 8, $\left(\frac{1}{2} \cdot 8\right)^2$ is 4^2, or 16. Thus, the perfect square $x^2 + 8x + 16$ represents the total area of square.

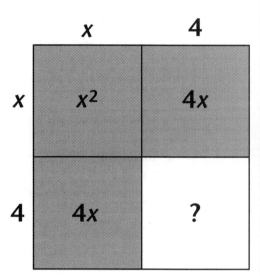

◆ **Example 1**
Complete the square for $x^2 - 4x$.

◆ **Solution**
The coefficient of x is 4. $\left(\frac{1}{2} \cdot 4\right)^2 = 2^2$, or 4

Thus, 4 completes the square.
The perfect square is $x^2 - 4x + 4$.

◆ **Example 2**
Find the minimum value for the function $f(x) = x^2 - 8x$.

◆ **Solution**
If a quadratic function is written in the form $f(x) = (x - h)^2 + k$, the minimum value of the function is k. To complete the square for $x^2 - 8x$, 16 must be added to the expression. This time, add and subtract **16**. The result is as follows:
$$f(x) = (x^2 - 8x + \mathbf{16}) - \mathbf{16}$$
$$= (x - 4)^2 - \mathbf{16}$$
The minimum value of f is -16.

Complete the square and find the minimum value of each function.

1. $f(x) = x^2 + 12x$

2. $g(x) = x^2 + 5x$

3. $h(x) = x^2 - 20x$

_____   _____   _____

4. $f(x) = x^2 - 2x$

5. $g(x) = x^2 - 10x$

6. $h(x) = x^2 + 11x$

_____   _____   _____

◆ **Skill B**   Finding the vertices of parabolas

Recall   If a quadratic function is written in the form $f(x) = (x - h)^2 + k$, the vertex of the parabola is $V(h, k)$. To find the vertex of the graph of $y = x^2 - 2x - 3$, complete the following steps:

- group the x^2 and the x term together. $y = (x^2 - 2x) - 3$
- complete the square. $\left(\frac{1}{2} \cdot 2\right)^2 = \mathbf{1}$
- add 1 and subtract 1. $y = (x^2 - 2x + \mathbf{1}) - 3 - \mathbf{1}$
- write -4 in the form $y = (x - h)^2 + k$. $y = (x - 1)^2 - 4$

The vertex of the parabola is $(1, 4)$.

◆ **Example**
Rewrite the equation $y = x^2 + 3x - 2$ in the form $y = (x - h)^2 + k$. The vertex is (h, k).

◆ **Solution**

$y = x^2 + 3x - 2$ The coefficient of x is 3.

$\quad = \left(x^2 + 3x + \frac{9}{4}\right) - 2 - \frac{9}{4}$ $\left(\frac{1}{2} \cdot 3\right)^2 = \left(\frac{3}{2}\right)^2 = \frac{9}{4}$

$\quad = \left(x + \frac{3}{2}\right)^2 - \frac{17}{4}$

The vertex is $\left(-\frac{3}{2}, -\frac{17}{4}\right)$.

Rewrite each equation in the form $y = (x - h)^2 + k$. Find the vertex of each parabola.

7. $y = x^2 + 2x + 1$ **8.** $y = x^2 - 8x + 3$ **9.** $y = x^2 + 4x - 3$

_____   _____   _____

10. $y = x^2 - 2x + 4$ **11.** $y = x^2 - 12x - 36$ **12.** $y = x^2 + 2x - 4$

_____   _____   _____

13. $y = x^2 - 3x + 6$ **14.** $y = x^2 - 5x - 25$ **15.** $y = x^2 + x + 1$

_____   _____   _____

Reteaching

8.4 Solving Equations of the Form $x^2 + bx + c = 0$

◆ **Skill A** Finding zeros of functions by completing the square

Recall The zeros of the function $f(x) = x^2 + bx + c$ are the values for which $f(x)$ is zero. One way to find the zeros of a function such as $g(x) = x^2 - 4x - 12$ is to set the expression equal to zero and solve the equation by completing the square.

$$x^2 - 4x - 12 = 0$$
$$x^2 - 4x = 12$$
$$x^2 - 4x + 4 = 12 + 4$$
$$(x - 2)^2 = 16$$
$$x - 2 = \pm 4$$
$$x = 2 + 4 \text{ or } x = 2 - 4$$
$$x = 6 \quad \text{or} \quad x = -2$$

The zeros are 6 and -2.

◆ **Example**
Solve the equation $x^2 + 6x - 6 = 0$ by completing the square.

◆ **Solution**

$$x^2 + 6x - 6 = 0$$
$$x^2 + 6x = 6$$
$$x^2 + 6x + 9 = 15$$
$$(x + 3)^2 = 15$$
$$x + 3 = \pm\sqrt{15}$$
$$x = -3 + \sqrt{15} \text{ or } x = -3 - \sqrt{15}$$

The solutions are $-3 + \sqrt{15}$ and $-3 - \sqrt{15}$.

Solve each equation by completing the square.

1. $y = x^2 + x - 2$ **2.** $y = x^2 + 3x - 10$ **3.** $y = x^2 + 5x + 6$

_____   _____   _____

4. $y = x^2 + 2x + 1$ **5.** $y = x^2 - 10x + 3$ **6.** $y = x^2 + 4x - 3$

_____   _____   _____

◆ **Skill B** Finding the zeros of functions by factoring

Recall The equation $x^2 - 4x - 12 = 0$ can also be solved by factoring the expression
$x^2 - 4x - 12$ and applying the Zero Product Property. The Zero Product Property
states that if $ab = 0$, then $a = 0$ or $b = 0$.
$$x^2 - 4x - 12 = 0$$
$$(x + 2)(x - 6) = 0$$
$$x + 2 = 0 \text{ or } x - 6 = 0$$
$$x = -2 \text{ or } x = 6$$
The solutions are -2 and 6.

◆ **Example 1**
Solve the equation $x^2 + 2x + 1 = 0$ by factoring.

◆ **Solution**
$$x^2 + 2x + 1 = 0$$
$$(x + 1)(x + 1) = 0$$
$$x + 1 = 0 \text{ or } x + 1 = 0$$
The only solution is -1.

◆ **Example 2**
Let $f(x) = x^2 - 4x + 3$. Find the value of x when $f(x)$ is 7.

◆ **Solution**
Solve $x^2 - 4x + 3 = 7$ by factoring or by completing the square.
$$x^2 - 4x + 3 = 7$$
$$x^2 - 4x = 4$$
$$x^2 - 4x + 4 = 4 + 4$$
$$(x - 2)^2 = 8$$
$$x - 2 = \pm\sqrt{8}$$
The solutions are $2 + \sqrt{8}$ and $2 - \sqrt{8}$

Solve each equation by factoring.

7. $y = x^2 - x - 20$ **8.** $y = x^2 - 7x + 12$ **9.** $y = x^2 + 4x - 5$

_____ _____ _____

Let $f(x) = x^2 + 6x - 2$. Find the value of x for each value of $f(x)$.

10. $f(x) = -10$ **11.** $f(x) = -2$ **12.** $f(x) = 0$

_____ _____ _____

Reteaching
8.5 The Quadratic Formula

◆ **Skill A**   Using the quadratic formula to solve equations

Recall   The solutions for a quadratic equation written in the form $ax^2 + bx + c = 0$, where $a \neq 0$, can be found by using the quadratic formula.

$$x = \frac{-b \pm \sqrt{b^2 - 4ac}}{2a}$$

◆ **Example**
Use the quadratic formula to solve $x^2 - 8x + 15 = 0$ for x.

◆ **Solution**
For $x^2 - 8x + 15 = 0$, a is 1, b is -8, and c is 15. Substitute these values in the quadratic formula.

$$x = \frac{-(-8) \pm \sqrt{(-8)^2 - (4)(1)(15)}}{(2)(1)}$$

$$= \frac{8 \pm \sqrt{64 - 60}}{2}$$

$$= \frac{8 \pm \sqrt{4}}{2}$$

$$= \frac{8 \pm 2}{2}$$

$$x = 3 \text{ or } x = 5$$

The solutions are 3 and 5.

Use the quadratic formula to solve each quadratic equation.

1. $x^2 - 5x + 4 = 0$

2. $x^2 - 2x - 24 = 0$

3. $x^2 + 6x + 9 = 0$

_____   _____   _____

4. $x^2 + 3x - 10 = 0$

5. $2x^2 - x - 6 = 0$

6. $2x^2 + x - 4 = 0$

_____   _____   _____

◆ **Skill B**   Using the quadratic formula to find the zeros of quadratic functions

Recall   The zeros of a quadratic function written in the form $f(x) = ax^2 + bx + c = 0$, where $a \neq 0$, can be found by using the quadratic formula.

◆ **Example**
Use the quadratic formula to find the zeros of $f(x) = 2x^2 - 5x - 3$.

◆ **Solution**
For $2x^2 - 5x - 1$, a is 2, b is -5, and c is -3. Substitute these values in the quadratic formula.

$$\frac{-(-5) \pm \sqrt{(-5)^2 - (4)(2)(-3)}}{(2)(2)} = \frac{5 \pm \sqrt{25 + 24}}{4} = \frac{5 \pm \sqrt{49}}{4} = \frac{5 \pm 7}{4}$$

The zeros are $-\frac{1}{2}$ and 3.

Use the quadratic formula to find the zeros of each quadratic function.

7. $f(x) = x^2 + 2x - 8$ **8.** $g(x) = 2x^2 - x - 15$ **9.** $h(x) = 4x^2 - 8x + 3$

_____   _____   _____

◆ **Skill C**   Using the discriminant to determine the number of solutions

Recall   When a quadratic equation is written in the form $ax^2 + bx + c = 0$, where $a \neq 0$, the expression $b^2 - 4ac$ is called the discriminant of the quadratic formula.
If $b^2 - 4ac > 0$, there are two solutions to the quadratic.
If $b^2 - 4ac = 0$, there is one solution to the quadratic.
If $b^2 - 4ac < 0$, there are no real number solutions.

◆ **Example**
What does the discriminant tell you about $3x^2 - 2x + 9 = 0$?

◆ **Solution**
For $3x^2 - 2x + 9 = 0$, a is 3, b is -2, and c is 9.
Thus, $b^2 - 4ac = (-2)^2 - (4)(3)(9) = 4 - 108 = -104$
The quadratic $3x^2 - 2x + 9 = 0$ has no real solutions.

Give the value of each discriminant. What does the discrimant tell you about the function?

10. $y = 4x^2 + 4x + 1$ **11.** $y = x^2 + 5x + 4$ **12.** $y = x^2 + 5x + 8$

_____   _____   _____

Reteaching
8.6 Graphing Quadratic Inequalities

◆ **Skill A** Solving quadratic inequalities

Recall The solution to a quadratic inequality such as $x^2 + 3x - 10 \leq 0$ can be graphed on a number line. First find the solution to $x^2 + 3x - 10 = 0$ by factoring, by completing the square, or by using the quadratic formula. Since $x^2 + 3x - 10 = (x + 5)(x - 2)$, the solutions to $x^2 + 3x - 10 = 0$ are -5 and 2. The points representing -5 and 2 divide the number line into three regions.

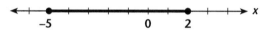

The region between -5 and 2 represents the solutions that are less than zero. The regions to the left of -5 and to the right of 2 represent the solutions that are greater than zero. Thus, the solution for the inequality is a closed line segment drawn between -5 and 2. The segment is closed because the inequality contains the equal sign.

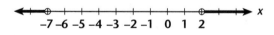

◆ **Example**
  Solve the quadratic inequality $x^2 + 5x - 14 > 0$. Graph the solution on a number line.

◆ **Solution**
  First solve the equality $x^2 + 5x - 14 = 0$.
  $x^2 + 5x - 14 = 0$
  $(x + 7)(x - 2) = 0$
  The solutions to the equality are -7 and 2. The regions to the left of -7 and to the right of 2 are the solutions to $x^2 + 5x - 14 > 0$. Since the equal sign is not part of the inequality, the graph will be two open rays.

```
←⊕─┼─┼─┼─┼─┼─┼─┼─┼─┼─⊕─┼→ x
  -7 -6 -5 -4 -3 -2 -1  0  1  2
```

Solve each quadratic inequality. Graph the solution on the number line.

1. $x^2 + 3x + 2 < 0$ _____

2. $x^2 - 8x + 15 \geq 0$ _____

3. $x^2 - 9 > 0$ _____

```
←┼┼┼┼┼┼┼┼┼┼┼┼┼┼┼→ x
```

4. $x^2 + x - 6 \leq 0$ _____

```
←┼┼┼┼┼┼┼┼┼┼┼┼┼┼┼→ x
```

NAME _____ CLASS _____ DATE _____

◆ Skill B Graphing quadratic inequalities

Recall To graph a quadratic inequality such as $y \leq x^2 + 2x - 8$, first graph the equality $y = x^2 + 2x - 8$. The solution to an inequality containing $<$ consists of all the points below the parabola. The solution to an inequality containing $>$ consists of all points above the parabola. If the inequality contains $\leq$ or $\geq$, the parabola is drawn with a solid line. Otherwise the parabola is drawn with a dashed line. The graph of $y \leq x^2 + 2x - 8$ is shown.

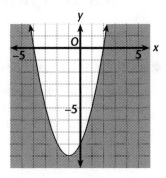

◆ Example
Graph the quadratic inequality $y \geq x^2 - 2x + 1$. Shade the solution region.

◆ Solution
First graph the quadratic inequality. Draw the parabola with a solid line. To be sure which region to shade, test the point $(0, 0)$ in the inequality. Since $0 \geq 1$ is false, shade the region that does not contain the point $(0, 0)$.

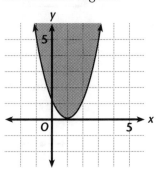

Graph each quadratic inequality. Shade the solution region for each inequality.

5. $y > x^2 - 6x$

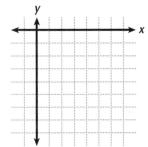

6. $y < x^2 + 4x + 4$

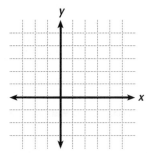

7. $y \geq x^2 - 4x + 5$

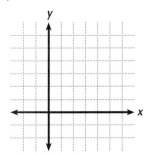

8. $y \leq x^2 - 3x + 2$

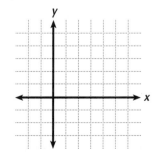

Reteaching
9.1 Exploring Square-Root Functions

◆ **Skill A** Finding square roots

Recall The number 25 is a perfect square because 25 objects can be arranged in a square with 5 objects on each side. The square root of 25 is the length of the side of the square. The square root of 25, written $\sqrt{25}$, is equal to 5. The solution of $x^2 = 25$ is $\pm\sqrt{25}$.

$\sqrt{25} = 5$ and $-\sqrt{25} = -5$

◆ **Example 1**
Find the square root of 4.84.

◆ **Solution**
$\sqrt{4.84} = 2.2$ because $2.2 \cdot 2.2 = 4.84$

◆ **Example 2**
Find the negative square root of 250.

◆ **Solution**
$\sqrt{250}$ is not a perfect square. The square root of 250 is an irrational number. The square root of 250 can only be approximated.
$-\sqrt{250} \approx -15.81$

Estimate each square root to the nearest hundredth.

1. $\sqrt{18}$

2. $\sqrt{115}$

3. $-\sqrt{9.6}$

_____

_____

_____

4. $\sqrt{134}$

5. $-\sqrt{0.08}$

6. $\sqrt{1500}$

_____

_____

_____

Find the length of the side of a square with each area.

7. 1369 square feet

8. 243.36 square meters

9. 100,000 square feet

_____

_____

_____

◆ **Skill B** Graphing square-root functions

Recall The function $f(x) = \sqrt{x}$, where $x \geq 0$, is the parent function for square roots. The function $g(x) = -\sqrt{x}$, where $x \geq 0$, reflects the square-root function through the x-axis. Taken together, however, $h(x) = \pm\sqrt{x}$ is not a function.

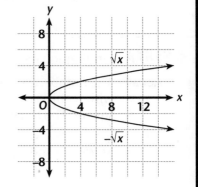

◆ **Example**
Graph $y = \sqrt{x + 3}$. Describe the transformation in terms of the square-root parent function.

◆ **Solution**

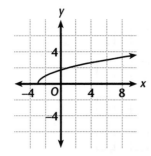

The graph of the parent function has been shifted horizontally to the left 3 units.

Graph each function. Describe each transformation in terms of the parent function.

10. $y = \sqrt{x} + 4$

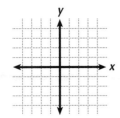

_____

11. $y = 2\sqrt{x} - 1$

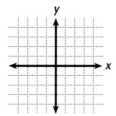

_____

12. $y = -\sqrt{x + 2}$

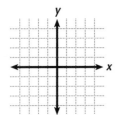

_____

13. $y = \sqrt{x - 3} + 2$

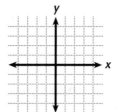

_____

 **Reteaching**
9.2 Operations With Radicals

◆ **Skill A**   Simplifying radicals

Recall   Radical expressions are in simplest radical form if the expression under the radical sign contains no perfect squares and there are no radicals in the denominator.

◆ **Example**
Simplify each of the following:

    a. $\sqrt{18}$ **b.** $\dfrac{\sqrt{4c^3}}{\sqrt{50c^2}}$

◆ **Solution**
    a. $\sqrt{18} = \sqrt{9 \cdot 2}\,\sqrt{3^2 \cdot 2} = \sqrt{3^2} \cdot \sqrt{2} = 3\sqrt{2}$

    b. $\dfrac{\sqrt{4c^3}}{\sqrt{50c^2}} = \dfrac{\sqrt{2^2 \cdot c^2 \cdot c}}{\sqrt{5^2 \cdot 2 \cdot c^2}} = \dfrac{2c\sqrt{c}}{5c\sqrt{2}} = \dfrac{2\sqrt{c}}{5\sqrt{2}}$

To further simplify the expression, multiply the denominator by a radical that will make a perfect square.

$$\frac{2\sqrt{c}}{5\sqrt{2}} = \frac{2\sqrt{c}}{5\sqrt{2}} \cdot \frac{\sqrt{2}}{\sqrt{2}} = \frac{2\sqrt{2}\sqrt{c}}{5 \cdot \sqrt{4}} = \frac{2\sqrt{2}\sqrt{c}}{5 \cdot 2} = \frac{\sqrt{2}\sqrt{c}}{5} = \frac{\sqrt{2c}}{5}$$

Simplify each radical expression.

1. $\sqrt{48}$ _____

2. $\sqrt{200b^2}$ _____

3. $\dfrac{\sqrt{72c^2}}{\sqrt{36c^3}}$ _____

4. $\dfrac{\sqrt{12a^4}}{\sqrt{20a^2}}$ _____

◆ **Skill B**   Adding radicals

Recall   Radical expressions with like radicands can be added by using the Distributive Property.

◆ **Example**
Simplify $\sqrt{8} + \sqrt{18}$.

◆ **Solution**
$\sqrt{8} + \sqrt{18} = \sqrt{2 \cdot 4} + \sqrt{2 \cdot 9} = 2\sqrt{2} + 3\sqrt{2} = (2 + 3)\sqrt{2} = 5\sqrt{2}$

Simplify each radical expression.

5. $2\sqrt{6} + 5\sqrt{6}$ _____

6. $3\sqrt{3} + \sqrt{27}$ _____

7. $\sqrt{50} + \sqrt{98}$ _____

8. $3\sqrt{2} + \sqrt{8} + \sqrt{2}$ _____

◆ **Skill C** Subtracting radicals

Recall Radical expressions with like radicands can also be subtracted.

◆ **Example**
Simplify $3\sqrt{2} - \sqrt{2} - \sqrt{12}$.

◆ **Solution**
$$3\sqrt{2} - \sqrt{2} - \sqrt{12} = (3\sqrt{2} - \sqrt{2}) - 2\sqrt{3}$$
$$= ((3 - 1)\sqrt{2}) - 2\sqrt{3}$$
$$= 2\sqrt{2} - 2\sqrt{3}$$

Simplify each radical expression.

9. $3\sqrt{5} - 7\sqrt{5}$ _____

10. $5\sqrt{8} - \sqrt{72}$ _____

11. $8\sqrt{75} - 9\sqrt{3}$ _____

12. $\sqrt{6} - \sqrt{24} - \sqrt{36}$ _____

◆ **Skill D** Multiplying radicals

Recall Expressions containing radicals can be multiplied by using the Distributive Property or the FOIL method.

◆ **Example**
Simplify $(2 + \sqrt{3})(5 - \sqrt{3})$.

◆ **Solution**
$$(2 + \sqrt{3})(5 - \sqrt{3}) = 2 \cdot 5 - 2 \cdot \sqrt{3} + \sqrt{3} \cdot 5 - \sqrt{3} \cdot \sqrt{3}$$
$$= 10 - 2\sqrt{3} + 5\sqrt{3} - 3$$
$$= 7 + 3\sqrt{3}$$

Simplify each radical expression.

13. $\sqrt{2}(\sqrt{6} + 5)$

14. $\sqrt{3}(5 - \sqrt{12})$

_____

_____

15. $\sqrt{6}(\sqrt{3} + \sqrt{2})$

16. $(2 + \sqrt{5})(2 - \sqrt{5})$

_____

_____

17. $(4 + \sqrt{12})(3 - \sqrt{2})$

18. $(\sqrt{3} + \sqrt{5})^2$

_____

_____

Reteaching
9.3 Solving Radical Equations

◆ **Skill A** Solving equations that contain radicals

Recall To solve an equation such as $\sqrt{x+3} = 2$, square each side of the equation to eliminate the radical sign. Then solve and check.

◆ **Example 1**
Solve $\sqrt{x+3} = 2$.

◆ **Solution**
$(\sqrt{x+3})^2 = 2^2$
$x + 3 = 4$
$x = 1$
The solution is 1 because $\sqrt{1+3} = 2$.

◆ **Example 2**
Solve $\sqrt{2x+8} = x$.

◆ **Solution**
$(\sqrt{2x+8})^2 = x^2$
$2x + 8 = x^2$
$x^2 - 2x - 8 = 0$
$(x-4)(x+2) = 0$
$x = 4$ or $x = -2$

Substitute 4 and -2 in the original equation to check the solutions.
Since $\sqrt{2(4)+8} = \sqrt{8+8} = \sqrt{16} = 4$, 4 is a solution.
Since $\sqrt{2(-2)+8} = \sqrt{-4+8} = \sqrt{4} \neq -2$, -2 is not a solution.
Thus, 4 is the only real solution of $\sqrt{2x+8} = x$.

Solve each radical equation.

1. $\sqrt{x-1} = 3$

2. $\sqrt{2x+3} = 5$

3. $\sqrt{x-7} = 4$

4. $\sqrt{x+8} = 4$

5. $\sqrt{x+20} = x$

6. $\sqrt{4x-4} = x$

◆ **Skill B**   Using the square-root generalization

Recall An equation such as $x^2 + 6x + 9 = 49$ can be solved by using the square-root generalization. If $x^2 = k$, $x = \sqrt{k}$ or $x = -\sqrt{k}$.
Since the left side of the equation is a perfect square, the equation can be written in the form $(x + 3)^2 = 49$.

$(x + 3)^2 = 49$
$x + 3 = \pm\sqrt{49}$
$x + 3 = 7$ or $x + 3 = -7$
$x = 4$ or $x = -10$
The solutions are 4 and -10 because $(4 + 3)^2 = 49$ and $(-10 + 3)^2 = 49$.

◆ **Example**
Solve $x^2 - 2x + 1 = 9$.

◆ **Solution**
$x^2 - 2x + 1 = 9$
$(x - 1)^2 = 9$
$x - 1 = \pm\sqrt{9}$
$x - 1 = 3$ or $x - 1 = -3$
$x = 4$ or $x = -2$
The solutions are 4 and -2, which check in the original equation.

Solve each equation by any method. Simplify any radicals.

7. $x^2 = 240$

8. $\sqrt{x} + 3 = 8$

9. $\sqrt{5x - 1} = 3$

_____

_____

_____

10. $\sqrt{6 - x} = 2$

11. $x^2 + 2x + 1 = 16$

12. $(x + 3)^2 = 20$

_____

_____

_____

13. $\sqrt{2x + 1} = 5$

14. $\sqrt{2x} = x$

15. $\sqrt{x^2 - 3x + 6} = x + 1$

_____

_____

_____

Reteaching
9.4 The "Pythagorean" Right-Triangle Theorem

◆ **Skill A**   Using the "Pythagorean" Right-Triangle Theorem

Recall   In a right triangle, the square of the hypotenuse is
equal to the sum of the squares of the legs.

$$c^2 = a^2 + b^2$$
$$a^2 = c^2 - b^2$$
$$b^2 = c^2 = a^2$$

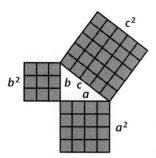

◆ **Example**
Find the hypotenuse of the right triangle.

◆ **Solution**
Let x equal the length of the hypotenuse.

$$x^2 = 4^2 + 4^2$$
$$x^2 = 32$$
$$x = \pm\sqrt{32}$$
$$x = \pm\sqrt{16 \cdot 2}$$
$$x = \pm4\sqrt{2}$$

Since the hypotenuse must be positive, the hypotenuse is $4\sqrt{2}$, or about 5.66.

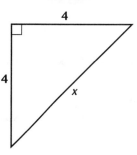

Find the missing length in each right triangle.

1.

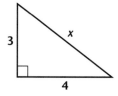

_____

2.

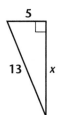

_____

3.

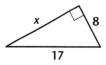

_____

4.

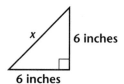

_____

5.

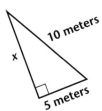

_____

6.

_____

◆ **Skill B**   Solving right-triangle problems

Recall   The "Pythagorean" Right-Triangle Theorem can be used to solve many problems involving right triangles.

◆ **Example**
The top of a 12-foot ladder reachs the top of a wall that is 10 feet tall. How far from the wall is the ladder?

◆ **Solution**
Draw a diagram. Note that the wall and the ground form a right angle. To find the distance, x, use the "Pythagorean" Right-Triangle Theorem.

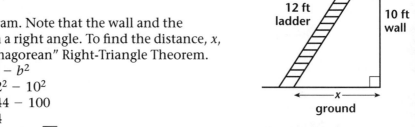

$$a^2 = c^2 - b^2$$
$$x^2 = 12^2 - 10^2$$
$$x^2 = 144 - 100$$
$$x^2 = 44$$
$$x = \pm\sqrt{44}, \text{ or } \pm2\sqrt{11}, \text{ or } \approx \pm6.6$$

Since the distance must be positive, the ladder is about 6.6 feet from the wall.

Solve each problem. Round the answer to the nearest hundredth, if necessary.

7. A ship leaves the marina and travels west for 45 miles. The ship then turns north and travels for 60 miles. How far is the ship from

the marina? _____

8. A rectangular swimming pool is 25 meters long and 12 meters wide.

What is the distance from corner to corner? _____

9. A square-shaped garden has a diagonal that measures 15 feet. What are

the dimensions of the garden? _____

 **Reteaching**
9.5 Exploring the Distance Formula

◆ **Skill A** Using the distance formula

Recall The distance between two points with coordinates $P(x_1, y_1)$ and $Q(x_2, y_2)$ is
$d = \sqrt{(x_2 - x_1)^2 + (y_2 - y_1)^2}$.

◆ **Example 1**
Use the distance formula to find the distance between $A(6, 9)$ and $B(-2, 5)$.

◆ **Solution**

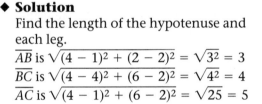

$$d = \sqrt{(-2 - 6)^2 + (5 - 9)^2}$$
$$= \sqrt{64 + 16}$$
$$= \sqrt{80}$$
$$= \sqrt{16 \cdot 5}$$
$$= 4\sqrt{5}$$

The distance between the points is $4\sqrt{5}$ units.

◆ **Example 2**
Use the distance formula and the "Pythagorean" Right-Triangle Theorem to prove that triangle ABC is a right triangle.

◆ **Solution**
Find the length of the hypotenuse and each leg.

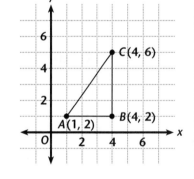

$\overline{AB}$ is $\sqrt{(4 - 1)^2 + (2 - 2)^2} = \sqrt{3^2} = 3$
$\overline{BC}$ is $\sqrt{(4 - 4)^2 + (6 - 2)^2} = \sqrt{4^2} = 4$
$\overline{AC}$ is $\sqrt{(4 - 1)^2 + (6 - 2)^2} = \sqrt{25} = 5$

$3^2 + 4^2 = 5^2$
Since the square of the hypotenuse is equal to the sum of the squares of the legs, triangle ABC is a right triangle.

Use the distance formula to find the distance between each pair of points.

1. $A(4, -3)$ and $B(-4, 3)$

2. $C(-2, -5)$ and $D(-2, 4)$

3. $E(7, 10)$ and $F(6, 2)$

_____ _____ _____

4. Points $L(-2, 3)$, $M(3, -2)$, and $N(6, 1)$ form a triangle. Prove that triangle LMN is or is not a right triangle.

◆ **Skill B** Using the midpoint formula

Recall The midpoint of a segment with endpoints $P(x_1, y_1)$ and $Q(x_2, y_2)$ is

$$M\left(\frac{x_1 + x_2}{2}, \frac{y_1 + y_2}{2}\right).$$

◆ **Example 1**
Find the midpoint of segment ST.

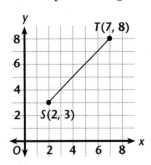

◆ **Solution**
Use the midpoint formula. Point S is $(2, 3)$ and point T is $(7, 8)$.

$$M\left(\frac{2 + 7}{2}, \frac{3 + 8}{2}\right) = M\left(\frac{9}{2}, \frac{11}{2}\right)$$

◆ **Example 2**
The midpoint of segment $\overline{AB}$ is $M(5, 7)$. One endpoint of the segment is $A(3, 4)$.
Find the coordinates of point $B(x_2, y_2)$.

◆ **Solution**
Use the midpoint formula to find x_2 and y_2.

$$5 = \frac{3 + x_2}{2} \qquad\qquad 7 = \frac{4 + y_2}{2}$$

$$3 + x^2 = 10 \qquad\qquad 4 + y_2 = 14$$

$$x_2 = 7 \qquad\qquad\quad y_2 = 10$$

The other endpoint is $B(7, 10)$.

Find the midpoint of each segment with the given endpoints.

5. $A(2, 1)$ and $B(-9, 3)$ **6.** $C(2, 2)$ and $D(-8, -7)$ **7.** $E(-1, -3)$ and $F(-3, -5)$

_____   _____   _____

8. The midpoint of a segement is $M(8, 2)$. One endpoint is $A(6, 0)$.

Find the coordinates of the other endpoint. _____

9. The endpoints of the diameter of a circle are $R(-8, 2)$ and $S(10, -4)$.

Find the coordinates of its center. _____

Reteaching
9.6 Exploring Geometric Properties

◆ **Skill A**   Finding the equations of circles with their centers at the origin

Recall   A circle is the set of all points in a plane that are the same distance from a given point called the center. Circle C has its center at the origin $(0, 0)$. Let r be the length of the radius of the circle and $P(x, y)$ be a point on the circle. Use the distance formula to find the equation of the circle.

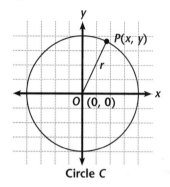

Circle C

$$(x - 0)^2 + (y - 0)^2 = r$$
$$x^2 + y^2 = r^2$$

The equation $x^2 + y^2 = r^2$ represents a circle with its center at the origin and a radius of r.

◆ **Example**
Find the equation of the circle with its center at the origin and a radius of 4.

◆ **Solution**
$x^2 + y^2 = 4^2$
The equation of the circle is $x^2 + y^2 = 16$.

Find the equation of each circle with its center at the origin and the given radius.

1. radius = 5 feet

2. radius = 2.5 meters

3. radius = $\frac{3}{4}$ inch

_____   _____   _____

◆ **Skill B**   Finding the equation of any circle

Recall   The equation of circle with center (h, k) and radius r is $(x - h)^2 + (y - k)^2 = r^2$.

◆ **Example**
Find the center and radius of the circle represented by the equation $(x - 2)^2 + (y + 3)^2 = 25$.

◆ **Solution**
h is 2, k is -3, and $r = \sqrt{25}$, or 5. The circle has its center at $(2, -3)$ and has a radius of 5.

Find the center and radius of each circle.

4. $(x + 5)^2 + (y + 1)^2 = 36$

5. $(x - 4)^2 + (y - 1)^2 = 27$

6. $(x - 3)^2 + (y - 3)^2 = 7.84$

_____   _____   _____

◆ **Skill C** Using coordinates to check geometric relationships

Recall Geometric figures can be described on a coordinate plane.

◆ **Example**
Check to see if the line segment drawn between the midpoints of the legs of an isosceles triangle is one-half of the length of it base.

◆ **Solution**
The midpoint formula gives (2, 3) for point A and (6, 3) for point B. The distance formula gives the length of $\overline{AB}$ as 4. The length of the base of the triangle is 8. Thus, the segment drawn between the midpoints is one-half of the length of the base.

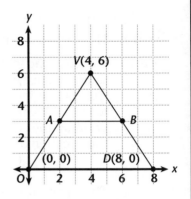

Draw each figure on a coordinate plane. Show that each relationship given is true.

7. A triangle is formed by the points $A(0, 0)$, $B(6, 0)$, and $C(3, 6)$. Show that two sides are equal.

8. A trapezoid is formed by the points $A(0, 0)$, $B(8, 0)$, $C(6, 4)$, and $D(2, 4)$. Show that the diagonals are equal.

9. A square is formed by the points $A(0, 0)$, $B(a, 0)$, $C(a, a)$, and $D(0, a)$. Show that the length of a diagonal equals $a\sqrt{2}$.

Reteaching
9.7 Tangent Function

◆**Skill A** Finding tangent ratios

Recall In a right triangle, the tangent of an acute angle is the ratio of the length of the leg opposite the angle to the length of the leg adjacent to the angle.

$$\tan \angle A = \frac{a}{b} \qquad \tan \angle B = \frac{b}{a}$$

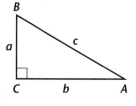

◆ **Example 1**
Find the tangent of $\angle M$ to the nearest thousandth.

◆ **Solution**

$$\tan \angle M = \frac{\text{opposite}}{\text{adjacent}} = \frac{1}{\sqrt{3}} = \frac{\sqrt{3}}{3} \approx 0.577$$

The tangent of angle M is about 0.577.

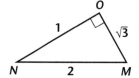

◆ **Example 2**
Use a calculator to find the tangent of 50° to the nearest thousandth.

◆ **Solution**
Use the tangent key on your calculator.
$\tan 50° \approx 1.192$

Find the tangent of each angle X to the nearest thousandth.

1.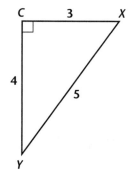

2.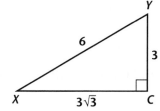

3.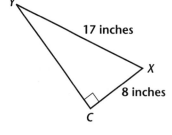

_____   _____   _____

Find the tangent of each angle to the nearest thousandth.

4. 20°

5. 45°

6. 72°

_____   _____   _____

◆ **Skill B**   Using the tangent ratio

Recall   The tangent ratio can be used to find a missing dimension in a right triangle.

◆ **Example**
The base of a pole is 20 feet from the stake at point A. The measure of the angle formed by the stake and the top of the pole is 55°. What is the height of the pole?

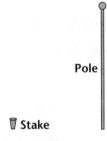

Pole

Stake

◆ **Solution**
Redraw the diagram to add the given information.

$$\tan \angle A = \frac{\text{opp}}{\text{adj}}$$

$$\tan 55° = \frac{a}{20}$$

$a = 20(\tan 55°)$
$a \approx (20)(1.428) \approx 28.56$
The height of the pole is about 28.56 feet.

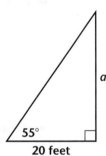

a

55°

20 feet

Use the diagrams to find each missing dimension to the nearest tenth.

7. Find the length of segment $\overline{YZ}$.

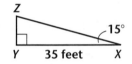

Z 15°
Y 35 feet X

_____

8. The measure of $\angle A$ is 25°. The length of segment AB is 50 meters. Find the length of segment $\overline{BC}$.

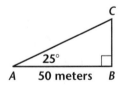

C
25°
A 50 meters B

_____

9. The height of a tree is 18 feet. Find the distance from the tree to the marker. The angle from the marker to the tree is 60°.

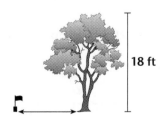

18 ft

_____

HRW Algebra One Interactions Course 2

Reteaching
9.8 Sine and Cosine Functions

◆ **Skill A**   Finding sines and cosines

Recall   The sine and cosine are ratios also based on a right triangle.

For triangle *ABC*:

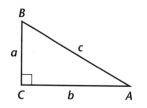

$$\sin A = \frac{a}{c} \text{ and } \cos A = \frac{b}{c}$$

$$\sin B = \frac{b}{c} \text{ and } \cos B = \frac{a}{c}$$

◆ **Example**
Find the sine and cosine of ∠*A*.

◆ **Solution**
Since *a* = 2 and *b* = 1, first use the "Pythagorean" Right-Triangle Theorem to find *c*.

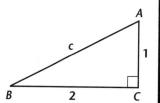

$$c = \sqrt{2^2 + 1^2} = \sqrt{5}$$

$$\sin A = \frac{2}{\sqrt{5}} = \frac{2\sqrt{5}}{5} \approx 0.894$$

$$\cos A = \frac{1}{\sqrt{5}} = \frac{\sqrt{5}}{5} \approx 0.447$$

Find the sine and cosine of each ∠A. Round to the nearest thousandth, if necessary.

1.
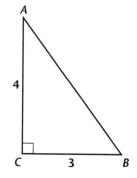

2.
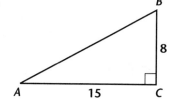

3.
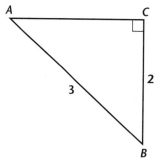

_____   _____   _____

Find the sine and cosine of each angle to the nearest thousandth.

4. 25°

5. 45°

6. 90°

_____   _____   _____

◆ **Skill B**   Using sine and cosine

Recall   The sine and cosine are ratios that can be used to find missing dimensions of a right triangle.

◆ **Example**
A 6-foot ramp is placed on the front end of a step. The ramp makes a 40° angle with the step. Find the height of the step.

◆ **Solution**
Draw a diagram to show the information.

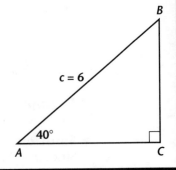

$\sin A = \dfrac{a}{c}$

$\sin 40° = \dfrac{a}{6}$

$a = 6(\sin 40°)$

$a \approx (6)(0.643) \approx 3.86$

The height of the step is about 3.86 feet.

Use a diagram to help find each missing dimension to the nearest tenth.

7. A kite is on a string 50 feet long. The kite string forms an angle of 35°

with the ground. How high above the ground is the kite? _____

8. Two miles of a railroad track is on a grade that forms a 15° angle with the horizontal. A 1500-meter train is on the tracks. How much higher

is the front of the train than the rear of the train? _____

9. A marker is placed 300 feet from the base of a hill. The marker makes a 20° angle with the top of the hill. What is the distance from the marker

to the top of the hill? _____

Reteaching
10.1 Rational Expressions

◆ **Skill A**   Evaluating rational expressions

Recall   A rational expression is an expression of the form $\frac{P}{Q}$, where P and Q are polynomials and $Q \neq 0$.

◆ **Example 1**
Identify the values for which the rational expression $\frac{x + 3}{x^2 - 4x + 4}$ is undefined.

◆ **Solution**
The rational expression is undefined when $x^2 - 4x + 4 = 0$.
Since $x^2 - 4x + 4 = 0$ when $(x - 2)^2 = 0$, the expression $x^2 - 4x + 4$ equals 0 when x is 2.
Thus, $\frac{x + 3}{x^2 - 4x + 4}$ is undefined when $x = 2$.

◆ **Example 2**
Evaluate the rational expression $\frac{x - 1}{x^2 - 4}$ for $x = 3$ and $x = 2$.

◆ **Solution**
For $x = 3$: $\frac{x - 1}{x^2 - 4} = \frac{3 - 1}{9 - 4} = \frac{2}{5}$

For $x = 2$: $\frac{x - 1}{x^2 - 4} = \frac{2 - 1}{4 - 4} = \frac{1}{0}$ Thus, $\frac{x - 1}{x^2 - 4}$ is undefined when $x = 2$.

Identify the values for which each rational expression is undefined.

1. $\frac{3x + 4}{x + 1}$

2. $\frac{5x}{x^2 + 5x - 6}$

3. $\frac{x^2 + 3x + 3}{x^2 - 9}$

_____   _____   _____

Evaluate each rational expression for $x = 3$ and $x = -1$.

4. $\frac{5x + 2}{3x - 12}$

5. $\frac{x + 3}{x^2 + 6x + 9}$

6. $\frac{x^2 - x - 2}{x^2 + 4x + 3}$

_____   _____   _____

◆ **Skill B**   Transforming rational expressions

Recall   The parent function for a rational function is $f(x) = \frac{1}{x}$, where $x \neq 0$.

◆ **Example**

Describe the transformations applied to the parent function $f(x) = \frac{1}{x}$ in order to produce the graph of $g(x) = \frac{1}{x} + 1$.

◆ **Solution**

Graph $f(x)$ and $g(x)$.

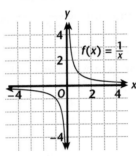   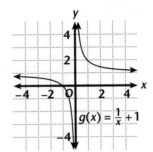

$$f(x) = \frac{1}{x} \qquad\qquad\qquad g(x) = \frac{1}{x} + 1$$

The graph of $f(x)$ has been shifted 1 unit up to produce the graph of $g(x)$.

Describe the transformations applied to the parent function $f(x) = \frac{1}{x}$, in order to produce the graph of each rational function.

7. $g(x) = \frac{1}{x} - 3$ _____

8. $g(x) = \frac{1}{x - 1}$ _____

9. $g(x) = \frac{1}{x + 1} + 2$ _____

10. $g(x) = \frac{2}{x + 2}$ _____

11. $g(x) = \frac{3}{x + 5} + 2$ _____

12. $g(x) = \frac{1}{x - 3} - 5$ _____

Reteaching
10.2 Inverse Variation

◆ **Skill A** Solving inverse variation equations

Recall If $\frac{k}{x} = y$ or $k = xy$, then y varies inversely with respect to x; that is, as x increases, y decreases and as x decreases, y increases for x, y, and $k > 0$. The constant of variation is the value of k. The graph of an inverse variation equation is a hyperbola.

◆ **Example 1**
Determine whether the equation $y = 2x$ describes inverse variation.

◆ **Solution**
In the equation $y = 2x$, as x increases, y also increases and its graph is a straight line. The equation $y = 2x$ does not describe inverse variation.

◆ **Example 2**
Suppose that y varies inversely as x. If y is 10 when x is 6, find y when x is 4.

◆ **Solution**
Since y varies inversely as x, $k = xy$.
$k = (6)(10)$ or 60
Since the constant of variation is 60, $4y$ must also equal 60.
If $4y = 60$, y must equal 15.

Determine whether each equation describes inverse variation.

1. $dt = 150$

2. $p = \frac{-10}{q}$

3. $2a - 5 = b$

_____ _____ _____

Suppose that y varies inversely as x. For each problem, if y is

4. 6 when x is 12,
find x when y is 9.

5. 25 when x is 4,
find x when y is 10.

6. $\frac{2}{3}$ when x is 15,
find x when y is 2.

_____ _____ _____

◆ **Skill B** Solving inverse variation problems

Recall If an equation involves inverse variation, find the constant of variation and then find the unknown quantity.

◆ **Example**
Time traveled varies inversely as the rate of travel. You can drive 8 hours at 50 miles per hour. How many hours will it take to make the same trip at 40 miles per hour?

◆ **Solution**
Distance is the constant of variation. Since distance is equal to the product of rate and time, $d = rt$ is the inverse variation equation to be solved.
$d = (8)(50) = 400$
The constant of variation is 400.
Thus, $40t = 400$ and $t = 10$.
It will take 10 hours to travel 400 miles at 40 miles per hour.

Solve each problem.

7. The base of a triangle with constant area varies inversely as its height. When the base is 36 inches, the height is 8 inches. Find the length of the base when the height is 12 inches.

_____

8. The current in an electric circuit varies inversely as the resistance. When the current is 60 amps, the resistance is 12 ohms. Find the current when the resistance is 16 ohms.

_____

9. The number of days to finish a job varies inversely as the number of people completing the job. If it takes 8 people 16 days to finish the job, how long will it take 12 people to finish the same job?

_____

10. The length of a rectangle with constant area varies inversely as its width. When the length is 10 meters, the width is 5 meters. Find the length when the width is 2 meters.

_____

Reteaching
10.3 Simplifying Rational Expressions

◆ Skill A   Factoring out monomials

Recall   A rational expression is an expression of the form $\frac{P}{Q}$, where P and Q are polynomials and $Q \neq 0$. A rational expression is in simplest form when the numerator and denominator have no common factors other than 1 or -1. Be sure to check for undefined terms before simplifying the expression.

◆ Example 1
Simplify the rational expression $\frac{6c - 9}{12c}$.

◆ Solution
First factor the numerator and denominator.

$$\frac{6c - 9}{12c} = \frac{3(2c - 3)}{(2)(2)(3)c}$$

Rewrite to show a fraction equal to 1.

$$= \frac{3}{3} \cdot \frac{2c - 3}{4c}$$

The simplified form is $\frac{2c - 3}{4c}$, where $c \neq 0$.

◆ Example 2
Simplify the rational expression $\frac{5m^2}{5m + 10m^2}$.

◆ Solution
First factor the numerator and denominator.

$$\frac{5m^2}{5m + 10m^2} = \frac{5 \cdot m \cdot m}{5 \cdot m \cdot (1 + 2m)}$$

Rewrite to show a fraction equal to 1.

$$= \frac{5m}{5m} \cdot \frac{m}{2m + 1}$$

The simplified form is $\frac{m}{2m + 1}$, where $m \neq 0$ and $m \neq -\frac{1}{2}$.

Simplify each rational expression. Give restrictions for each variable.

1. $\frac{14t}{7t - 7}$

2. $\frac{3m + 9}{6m - 12}$

3. $\frac{4m^2}{8m + 12m^2}$

4. $\frac{3y + 6}{y^2 + 4y + 2}$

5. $\frac{3x + 3}{x^2 + 2x + 1}$

6. $\frac{4r - 12}{r^2 - 6r + 9}$

◆ **Skill B**   Factoring out binomials

Recall   A rational expression is in simplest form when the numerator and denominator have no common factors other than 1 or -1.

◆ **Example**
Simplify the rational expression $\dfrac{x-1}{x^2-3x+2}$.

◆ **Solution**
First factor the numerator and denominator. $\dfrac{x-1}{x^2-3x+2}=\dfrac{x-1}{(x-1)(x-2)}$

Write the restrictions on the variable. $x\neq1$ and $x\neq2$

Rewrite to show a fraction equal to 1. $=\dfrac{x-1}{x-1}\cdot\dfrac{1}{x-2}$

The simplified form is $\dfrac{1}{x-2}$, where $x\neq1$ and $x\neq2$.

Simplify each rational expression. Give restrictions for each variable.

7. $\dfrac{m^2+3m+2}{m+1}$

8. $\dfrac{a-2}{a^2-5a+6}$

9. $\dfrac{3b+6}{b^2+4b+4}$

10. $\dfrac{r^2-r}{r^2-1}$

11. $\dfrac{x^2-1}{x^2-x}$

12. $\dfrac{x^2+x-6}{x^2-6x+8}$

13. The area of a rectangular flower bed is represented by $x^2+3x-10$. One side of the flower bed is represented by $x+5$. What is the length of the other side of the flower bed?

NAME _____ CLASS _____ DATE _____

 **Reteaching**
10.4 Operations with Rational Expressions

◆ **Skill A** Multiplying rational expressions

Recall To multiply two rational expressions, multiply the numerators and multiply the denominators. Then simplify the results. List any restrictions.

◆ **Example**

Multiply $\dfrac{3a + 3}{a} \cdot \dfrac{a - 3}{a + 1}$.

◆ **Solution**

Multiply numerators and denominators. $\dfrac{3a + 3}{a} \cdot \dfrac{a - 3}{a + 1} = \dfrac{(3a + 3)(a - 3)}{a(a + 1)}$

Factor all expressions. $= \dfrac{3(a + 1)(a - 3)}{a(a + 1)}$

Rewrite to show a fraction equal to 1. $= \dfrac{a + 1}{a + 1} \cdot \dfrac{3(a - 3)}{a}$

Simplify and note restrictions. $= \dfrac{3(a - 3)}{a}$,

where $a \neq 0$ and $a \neq -1$

Find each product.

1. $\dfrac{2t - 2}{t} \cdot \dfrac{t^2}{t - 1}$

2. $\dfrac{12a - 18}{18a} \cdot \dfrac{2a}{2a - 3}$

3. $\dfrac{5}{c + 3} \cdot \dfrac{c^2 + 6c + 9}{10}$

4. $\dfrac{20d}{d^2 + 8d + 15} \cdot \dfrac{d + 3}{5d}$

5. $\dfrac{x^2 - 9}{4} \cdot \dfrac{8}{x + 3}$

6. $\dfrac{y^2 - y - 2}{y + 2} \cdot \dfrac{y^2 + y - 2}{y - 2}$

◆ **Skill B**   Adding and subtracting rational expressions

Recall   Rewrite all rational expressions with common denominators. Add or subtract the numerators. Keep the same denominator. Then simplify the results. List any restrictions.

◆ **Example**

Subtract $\dfrac{b}{b+3} - \dfrac{4}{b+2}$.

◆ **Solution**

Multiply numerators and denominators by fractions equal to 1 that will result in common denominators.

$$\dfrac{b+2}{b+2} \cdot \dfrac{b}{b+3} - \dfrac{b+3}{b+3} \cdot \dfrac{4}{b+2}$$

Multiply the rational expressions.

$$\dfrac{b^2+2b}{(b+2)(b+3)} - \dfrac{4b+12}{(b+2)(b+3)}$$

Subtract and simplify.

$$\dfrac{b^2-2b-12}{(b+2)(b-3)}$$

Find each sum or difference.

7. $\dfrac{x}{4} + \dfrac{2x}{5}$

8. $\dfrac{7}{2b} + \dfrac{8}{3b}$

_____

9. $\dfrac{t}{4t^2} - \dfrac{5}{6t}$

10. $\dfrac{2m}{m-1} + \dfrac{m}{2m-2}$

_____

11. $\dfrac{3c}{3c-12} - \dfrac{c}{2c-8}$

12. $\dfrac{y}{y+2} + \dfrac{2y}{y-2}$

_____

Reteaching
10.5 Solving Rational Equations

◆ **Skill A** Using the common denominator method to solve rational equations

Recall To solve an equation containing rational expressions, first multiply each side of the equality by the common denominator of all the rational expressions. This will clear the equations of fractions. Solve the resulting equation. Check the solutions and the restrictions.

◆ **Example**

Solve. **a.** $\dfrac{x}{3} + \dfrac{x+1}{4} = \dfrac{17}{12}$ **b.** $\dfrac{3}{x-1} - 2 = -1$

◆ **Solution**

a. Multiply each side of the equation by 12.

$$12\left(\dfrac{x}{3} + \dfrac{x+1}{4}\right) = 12 \cdot \dfrac{17}{12}$$

$$4x + 3x + 3 = 17$$

$$7x = 14$$

$$x = 2$$

b. Multiply each side of the equation by $x - 1$.

$$(x-1)\left(\dfrac{3}{x-1} - 2\right) = (x-1)(-1),\text{ where } x \neq 1$$

$$(x-1)\left(\dfrac{3}{x-1}\right) - (x-1)(2) = (x-1)(-1)$$

$$3 - 2x + 2 = -x - 1$$

$$x = 4$$

Solve each equation.

1. $\dfrac{a}{3} - \dfrac{a}{2} = 1$

2. $\dfrac{5}{d} + \dfrac{3}{d^2} = \dfrac{13}{4}$

3. $\dfrac{h-2}{h} - \dfrac{h-3}{h-6} = \dfrac{1}{h}$

4. $\dfrac{7}{x-4} - \dfrac{5}{x-2} = 0$

♦ **Skill B** Solving work problems by using rational equations

Recall Work problems deal with situations in which people work at different rates to complete a job. The formula usually used for work problems is rate of work · time = work done, or $r \cdot t = w$.

For example, suppose that it takes 5 hours to paint one room. Then $\frac{1}{5}$ of the job can be completed in 1 hour; $2 \cdot \frac{1}{5}$, or $\frac{2}{5}$, of the job can be completed in 2 hours; and $n \cdot \frac{1}{5}$, or $\frac{n}{5}$, of the job can be completed in n hours. Completing the whole job is $5 \cdot \frac{1}{5}$, or 1.

♦ **Example**
Paul can clean a room in 60 minutes. Pat can clean the same room in 30 minutes. If Paul and Pat work together, how long will it take to clean the room?

♦ **Solution**
Let t represent the number of minutes it will take working together.

Paul's rate of work is $\frac{1}{60}$ of the job per minute. His part is $\frac{t}{60}$.

Pat's rate of work is $\frac{1}{30}$ of the job per minute. Her part is $\frac{t}{30}$.

The total of the two parts of the job equals the whole job, or 1.

Thus,
$$\frac{t}{60} + \frac{t}{30} = 1$$
$$60\left(\frac{t}{60} + \frac{t}{30}\right) = 60 \cdot 1$$
$$t + 2t = 60$$
$$3t = 60$$
$$t = 20$$
Together, Paul and Pat can clean the room in 20 minutes.

Solve each problem.

5. It takes 4 hours to fill a swimming pool from one hose. It takes 6 hours with a smaller hose. How long will it take to fill the pool if both hoses are used at the same time?

6. Working together, Sergi and Anita can wash all of the windows in their house in 3 hours. Working by himself, Sergi can wash the windows in 5 hours. How long will it take Anita to wash the windows if she does the job alone?

Reteaching
10.6 Exploring Proportions

◆ **Skill A** Solving proportions

Recall A proportion is an equation which states that two ratios are equal. In a proportion, the product of the means equals the product of the extremes. In other words, cross-multiply.

$$\text{If } \frac{a}{b} = \frac{c}{d}, \text{ then } ad = bc, \text{ where } b \neq 0 \text{ and } d \neq 0$$

◆ **Example**
Solve the proportion $\frac{6}{x} = \frac{9}{x + 2}$.

◆ **Solution**
Cross-multiply.   If $\frac{6}{x} = \frac{9}{x + 2}$, then $6(x + 2) = 9 \cdot x$.

$$6x + 12 = 9x$$
$$3x = 12$$
$$x = 4$$

Also check the restrictions, $x \neq 0$ and $x \neq -2$. Since 4 is not one of the restricted values, 4 is the solution to the proportion.

Solve each proportion.

1. $\dfrac{5}{2a + 1} = \dfrac{5}{7}$

2. $\dfrac{b + 2}{b} = \dfrac{4}{5}$

3. $\dfrac{2}{3x + 5} = \dfrac{3}{2x}$

4. $\dfrac{3}{r - 3} = \dfrac{r - 3}{3}$

5. $\dfrac{2}{d - 5} = \dfrac{d}{3}$

6. $\dfrac{m + 2}{m - 5} = \dfrac{m}{m - 2}$

◆ **Skill B**   Using proportions to solve problems

Recall   Set up a proportional relationship and solve.

◆ **Example**
The area of a rectangular lot is 135 square yards. The sides of the lot are in a ratio of three to five. What are the dimensions of the lot.

◆ **Solution**
Let one side $= 3x$ and the other side $= 5x$. Then,
$3x \cdot 5x = 135$
$15x^2 = 135$
$x^2 = 9$
$x = 3$ (the dimensions must be positive)
$3x = 3(3) = 9$
$5x = 5(3) = 15$
The dimensions are 9 yards by 15 yards.

Use a proportion to solve each problem.

7. Mr. Ryan can drive his car 375 miles on 15 gallons of gasoline. How far can he drive his car on 12 gallons of gasoline?

_____

8. Two supplementary angles are in a ratio of 4 to 5. Find the measure of the smaller angle.

_____

9. It cost $25.50 for several friends to buy tickets to a movie. For $34.00, 2 more people could have bought tickets. How many friends bought tickets?

_____

10. Two complementary angles are in a ratio of 2 to 6. Find the measure of each angle.

_____

Reteaching
10.7 Proof in Algebra

♦**Skill A** Using proofs to solve equations

Recall A statement written in if-then form is called a conditional. The *if* clause is assumed to be true and the properties of algebra are used to show that the *then* clause follows logically.

♦ **Example**
Use the properties of algebra to prove the following conditional:

If $\dfrac{10}{x+1} = \dfrac{15}{12}$, then $x = 7$.

♦ **Solution**
Start with the *if* clause and give a reason for each step until you reach the *then* clause.

$\dfrac{10}{x+1} = \dfrac{15}{12}$	Given
$15(x+1) = 12(10)$	Cross products
$15x + 15 = 120$	Distributive Property
$15x = 105$	Subtraction Property
$x = 7$	Division Property

Use the properties of algebra to solve each equation. Give a reason for each step.

1. $3x - 6 = 15$

2. $8x^2 = 72$

3. $x^2 - x = 12$

4. $\dfrac{5}{x} = \dfrac{x-3}{2}$

◆ **Skill B** Proving theorems

Recall When a statement can be proven for every case, the conditional is called a theorem.

◆ **Example**
Use the properties of algebra to prove the following theorem:
The square of an odd integer is always an odd integer.

◆ **Solution**

Let x represent any integer.	Given
$2x$ is an even integer.	Any integer in the form $2x$ is even.
$2x + 1$ is odd.	Any odd integer is an even number plus 1.
$(2x + 1)^2 = 2x^2 + 4x + 1$	Squaring a binomial
$\qquad = 2(x^2 + 2x) + 1$	Distributive Property

The integer represented by $2(x^2 + 2x) + 1$ is odd because it is an even number, $2(x^2 + 2x)$, plus 1. Thus, any odd integer squared will always result in an odd integer.

Prove each theorem. Give a reason for each step.

5. The sum of two odd numbers is even.

6. The product of consecutive odd and even numbers is even.

7. The sum of consecutive integers is odd.

8. The square of the sum of two consecutive integers is odd.

9. The sum of two multiples of 3 is also a multiple of 3.

10. The product of an even number and a multiple of 5 is divisible by 10.

ANSWERS

Reteaching—Chapter 1

Lesson 1.1

1. 8 **2.** 19 **3.** 36 **4.** 63 **5.** 75 **6.** c

7. p **8.** m **9.** x **10.** t

11. $m + 22 - 22 = m$ **12.** $g + 5 - 5 = g$

13. $c - 25 + 25 = c$ **14.** $c + 15$, or $15 + c$

15. $t - 5$ **16.** $17 - p$ **17.** $j + 13$, or $13 + j$

18. $m = 34$ **19.** $y = -1.9$ **20.** $z = 13$

21. $p = 25$ **22.** $i = 5.0$ **23.** $c = 13$

24. $x = 15$ **25.** $r = 125$ **26.** $t = -6.02$

27. $r = 12$ **28.** $v = 73.8$ **29.** $y = 50$

30. $m = -51$ **31.** $p = -0.6$ **32.** $z = -14$

33. $c = 6$

Lesson 1.2

1. $-\frac{4}{3}$ **2.** $\frac{1}{3}$ **3.** 8 **4.** $-\frac{1}{7}$ **5.** $\frac{5}{8}$ **6.** $-\frac{2}{3}$

7. $y = -12$ **8.** $x = -4.5$ **9.** $w = \frac{3}{2}$, or 1.5

10. $240 = 0.80r$; $\$300 = r$

11. $180 = 0.90r$; $\$200 = r$ **12.** 840

Lesson 1.3

1. multiplication; subtraction

2. multiplication; addition

3. division; addition

4. division; subtraction

5. multiplication; addition

6. division; subtraction or addition

7. multiplication; subtraction

8. division; addition

9. multiplication; addition

10. division; subtraction

11. $p = 10$ **12.** $m = 24$ **13.** $r = 7$

14. $c = 6$ **15.** $y = 5.6$ **16.** $x = -42$

17. $t = 165$ **18.** $c = 30$ **19.** $r = 16$

20. $y = -96$ **21.** 4 **22.** 27 **23.** 1

24. 5 **25.** 4 **26.** -1 **27.** $w = 21$

28. $l = 20$ **29.** $l = 13$ **30.** $w = 20$

31. $s = 20$ feet **32.** $s = 12$ meters

Lesson 1.4

1. $8p$ **2.** $10m$ **3.** $15x^2$ **4.** $-y$ **5.** $16y^2$

6. $8n$ **7.** $2m^2 + 8m$ **8.** $-2t^2 + 8t - 1$

9. $19c^2 - 2c + 2$ **10.** $-6r^2 - 4r - 12$

11. $y = 3$ **12.** $x = -13$ **13.** $x = 2.75$

14. $w = 5.5$ **15.** $-y - 4$ **16.** $-c + 5$

17. $t - 2$ **18.** $m + 4$ **19.** $x - 7$

20. $4x + 23$ **21.** $x + 6$ **22.** $2x + 5$

23. $-3x - 2$ **24.** $2m - 8$ **25.** $x = 10$

26. $y = 3$ **27.** $p = -24$ **28.** $r = 25$

29. $h = 5$ **30.** $q = 5$ **31.** $x = 7$

32. $s = 1.5$ **33.** $t = 0$ **34.** $c = 2$

35. $g = -1$ **36.** $v = -0.4$

Lesson 1.5

1. $2n > 8$ **2.** $\frac{n}{5} \le -11$ **3.** $16n < 5$

4. $\frac{n}{13} \ge 52$

5. $2n \cdot 4 \le n + 8$, or $8n \le n + 8$

ANSWERS

6. $2x + 520 \geq 670$

7.
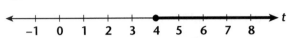

8.

9. $x \geq 7$ **10.** $t > -12$ **11.** $p < -11$

12. $y \leq 2$

Lesson 1.6

1. ± 12 **2.** ± 5 **3.** ± 24 **4.** ± 5.3

5. ± 11 **6.** ± 15 **7.** ± 14 **8.** ± 36

9. $x = 2.8; x = -0.6$ **10.** $c = 38; c = -53$

11. $y = 28; y = -16$ **12.** $x = 13; x = -23$

13. $r = 4.9; r = -2.7$ **14.** $p = 20; p = -38$

15. $q = 62; q = -78$ **16.** $a = 63; a = -61$

17. $a = 7.3; a = 2.7$

18. $x = -14.5; x = -29.5$

19. $t = -20; t = -30$ **20.** $p = 10; p = -1.4$

21. no solution   **22.** $q = -69; q = -75$

23. $|x - 13,200| \leq 100$; the guess must be between \$13,100 and \$13,300 inclusive.

24. $|x - 310| \leq 12$; between 298 and 322 minutes inclusive.

25. $|x - 1800| \leq 50$; between 1750 and 1850 calories inclusive.

Reteaching—Chapter 2

Lesson 2.1

1. negative   **2.** positive   **3.** undefined

4. $\frac{3}{2}$ **5.** $-\frac{3}{2}$ **6.** 0 **7.** $-\frac{4}{3}$ **8.** -1 **9.** $\frac{1}{6}$

10. $-\frac{1}{2}$ **11.** $\frac{9}{11}$ **12.** $-\frac{9}{13}$

Lesson 2.2

1. $m = -\frac{1}{3}; b = 6$ **2.** $m = 1; b = -1$

3. $m = -\frac{3}{5}; b = 3$ **4.** $m = -2; b = -5$

5. $m = \frac{1}{2}; b = -2$ **6.** $m = \frac{5}{2}; b = 2$

7. $m = \frac{2}{3}; b = 3$ **8.** $m = -\frac{1}{3}; b = -3$

9. $m = 3; b = 0$ **10.** $m = 0; b = -6$

11.
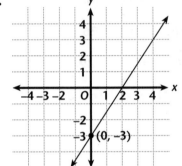

12.
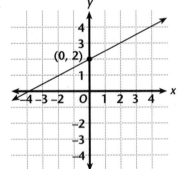

13.
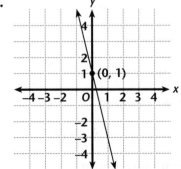

ANSWERS

14.

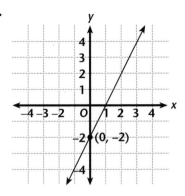

15.

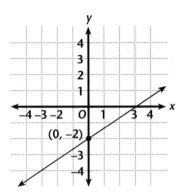

16.

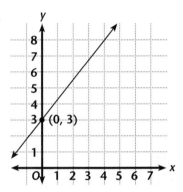

17. $y = \frac{3}{2}x - \frac{27}{2}$ **18.** $y = 3x + 12$

19. $y = 20x + 120$ **20.** $y = 2x - 20$

Lesson 2.3

1. yes   **2.** no   **3.** yes   **4.** yes

5. $x = \frac{1}{2}, y = 0$ **6.** $x = -2, y = -1$

7. $(3, 1)$

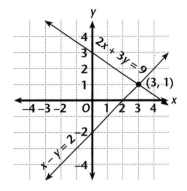

8. $(0, 4)$

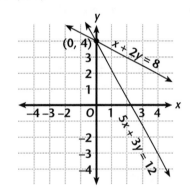

9. $(3, 6)$

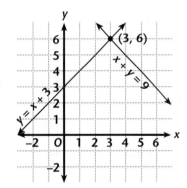

10. $(1, -2)$

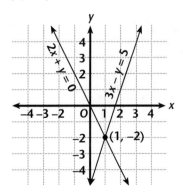

ANSWERS

Lesson 2.4

1. $x = 3$ **2.** $x = 1$ **3.** $y = -1$ **4.** $y = 2$

5. $x = -2$ **6.** $y = 3$ **7.** $x = 1$ **8.** $y = 1$

9. $x = 3, y = 6$ **10.** $x = 5, y = 1$

11. $x = 3, y = 5$ **12.** $x = 3, y = 6$

Lesson 2.5

1. $2t$ **2.** $-5p$ **3.** $-8c$ **4.** $3m$ **5.** 36

6. 15   **7.** 30   **8.** 20   **9.** 30   **10.** 24

11. 100   **12.** 48   **13.** $x = 2, y = 4$

14. $x = 4, y = 5$ **15.** $a = 5, b = 2$

16. $x = 4, y = 3$

Lesson 2.6

1. inconsistent   **2.** consistent

3. inconsistent   **4.** consistent

5. inconsistent   **6.** dependent

7. independent   **8.** dependent

9. one   **10.** none   **11.** infinite   **12.** one

Lesson 2.7

1. dashed   **2.** solid   **3.** solid   **4.** dashed

5.
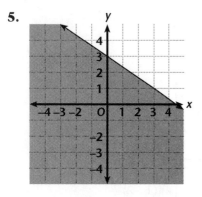

6.
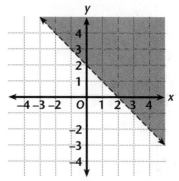

7.
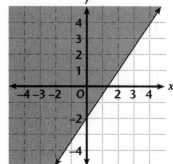

8.
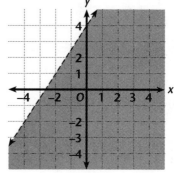

9.
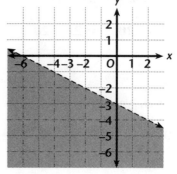

10.

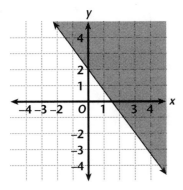

11.

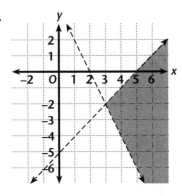

12.

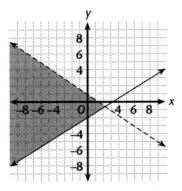

13.

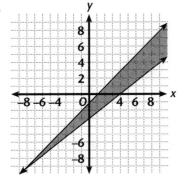

14.

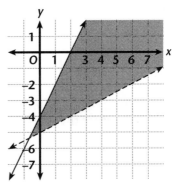

Lesson 2.8

1. Let x represent pounds of candy worth $1.35/lb.
Let y represent pounds of candy worth $1.95/lb.

$$\begin{cases} x + y = 80 \\ 1.35x + 1.95y = 1.50(80) \end{cases}$$

2. Let x represent the number of $5 bills.
Let y represent the number of $2 bills.

$$\begin{cases} x + y = 23 \\ 5x + 2y = 91 \end{cases}$$

3. Let x represent the tens digit.
Let y represent the ones digit.

$$\begin{cases} x + y = 9 \\ 10x + y = 6y \end{cases}$$

4. rowing rate in still water = 6 mph
rate of current = 2 mph

5. 6 years old

Reteaching—Chapter 3

Lesson 3.1

1. A: 3×3; B: 3×3; C: 3×2; D: 2×3

2. yes; $a_{22} = 4$; $b_{22} = \sqrt{16} = 4$

3. Yes; B simplifies to $\begin{bmatrix} 2 & 0 & -3 \\ -1 & 4 & 6 \\ 3 & -2 & 5 \end{bmatrix}$, which is the same as A.

ANSWERS

‒ ‒

4. No; they have different dimensions.

5. $x = -16, y = -3, z = -24$

6. $a = 1, b = 3, c = -5, d = 4, e = -2$

7. Substitute the values in the original matrices, and see whether the matrices are equal.

Lesson 3.2

1. -6 **2.** -25 **3.** 47 **4.** 7 **5.** -92

6. 52 **7.** 93 **8.** 21

9. $\begin{bmatrix} 4.7 & -7.9 & 15 \\ -0.8 & 0 & 1 \end{bmatrix}$

10. cannot be added **11.** $\begin{bmatrix} 2.5 & 3.5 \\ 0.5 & 7.5 \\ 7 & -5 \end{bmatrix}$

12. $\begin{bmatrix} 8 & -2 & -4 \\ -5 & 8 & -6 \end{bmatrix}$ **13.** $\begin{bmatrix} 2 & 14 \\ 6 & -2 \end{bmatrix}$

14. $\begin{bmatrix} 9 & -7 \\ 7 & -3 \\ 10 & -11 \end{bmatrix}$

Lesson 3.3

1. yes **2.** yes **3.** no **4.** no **5.** no

6. yes **7.** $\begin{bmatrix} -16 & -24 \\ 10 & 15 \end{bmatrix}$ **8.** $\begin{bmatrix} 11 \\ -8 \end{bmatrix}$

9. no product possible **10.** $[14]$

11. $\begin{bmatrix} 17 & -7 \\ -5 & 22 \end{bmatrix}$ **12.** $\begin{bmatrix} -3 & -2 & -1 \\ -6 & -4 & -2 \\ -9 & -6 & -3 \end{bmatrix}$

Lesson 3.4

1. $\begin{cases} 4a + 6c = 1 \\ -2a + 5c = 0 \end{cases} \begin{cases} 4b + 6d = 0 \\ -2b + 5d = 1 \end{cases}$

2. $\begin{cases} 2a - 6c = 1 \\ a + 3c = 0 \end{cases} \begin{cases} 2b - 6d = 0 \\ b + 3d = 1 \end{cases}$

3. $\begin{cases} 5a + 9c = 1 \\ -4a - 2c = 0 \end{cases} \begin{cases} 5b + 9d = 0 \\ -4b - 2d = 1 \end{cases}$

4. $\begin{cases} -a + 0c = 1 \\ 5a - 3c = 0 \end{cases} \begin{cases} -b + 0d = 0 \\ 5b - 3d = 1 \end{cases}$

5. $\begin{cases} a - 2c = 1 \\ 4a + 6c = 0 \end{cases} \begin{cases} b - 2d = 0 \\ 4b + 6d = 1 \end{cases}$

6. $\begin{cases} 9a - 5c = 1 \\ -2a + 4c = 0 \end{cases} \begin{cases} 9b - 5d = 0 \\ -2b + 4d = 1 \end{cases}$

7. $\begin{cases} -a + 4c = 1 \\ -2a + 5c = 0 \end{cases} \begin{cases} -b + 4d = 0 \\ -2b + 5d = 1 \end{cases}$

8. $\begin{cases} 5a - c = 1 \\ 0a - 2c = 0 \end{cases} \begin{cases} 5b - d = 0 \\ 0b - 2d = 1 \end{cases}$

9. $\begin{bmatrix} 3 & -4 \\ -2 & 3 \end{bmatrix}$ **10.** $\begin{bmatrix} 2 & -5 \\ -1 & 3 \end{bmatrix}$

11. $\begin{bmatrix} 5 & -17 \\ -2 & 7 \end{bmatrix}$ **12.** $\begin{bmatrix} 7 & -6 \\ -8 & 7 \end{bmatrix}$

Lesson 3.5

1. $\begin{bmatrix} 8 & 3 \\ -3 & 2 \end{bmatrix} \begin{bmatrix} x \\ y \end{bmatrix} = \begin{bmatrix} 4 \\ 11 \end{bmatrix}$

2. $\begin{bmatrix} 2 & -5 \\ 7 & 9 \end{bmatrix} \begin{bmatrix} x \\ y \end{bmatrix} = \begin{bmatrix} 16 \\ 3 \end{bmatrix}$

3. $\begin{bmatrix} 9 & 6 \\ 3 & 2 \end{bmatrix} \begin{bmatrix} x \\ y \end{bmatrix} = \begin{bmatrix} 15 \\ 5 \end{bmatrix}$

4. $\begin{bmatrix} -5 & 2 \\ 10 & -4 \end{bmatrix} \begin{bmatrix} x \\ y \end{bmatrix} = \begin{bmatrix} -3 \\ 6 \end{bmatrix}$

5. $\begin{bmatrix} -5 & 2 \\ 3 & -1 \end{bmatrix} \begin{bmatrix} x \\ y \end{bmatrix} = \begin{bmatrix} 3 \\ 1 \end{bmatrix}$

6. $\begin{bmatrix} 9 & -5 \\ -2 & 3 \end{bmatrix} \begin{bmatrix} x \\ y \end{bmatrix} = \begin{bmatrix} 4 \\ 1 \end{bmatrix}$

7. $x = 2, y = 3$ **8.** $x = -2, y = 1$

9. $x = 18, y = 20$ **10.** $x = \frac{1}{3}, y = -\frac{1}{2}$

Reteaching—Chapter 4

Lesson 4.1

1. $0.80; 80\%$ **2.** $0.625; 62.5\%$

3. $1.5; 150\%$

ANSWERS

4. $0.\overline{285714}$, or about 28.6%

5. $0.33\overline{3}$, or about 33.3%

6. $0.26\overline{6}$, or about 26.6%

7. $\frac{17}{100} = 17\%$ **8.** $\frac{19}{100} = 19\%$

9. $\frac{64}{100} = 64\%$ **10.** $\frac{83}{100} = 83\%$

11. $\frac{100}{100} = 100\%$

Lesson 4.2

1. Answers will vary. Possible answer: Let heads represent true and tails represent false. Let each toss represent one trial. Generate 24 tosses (trials). The number of tails generated represents the number of students that guessed correctly.

2. Answers will vary. Possible answer: Let heads represent an even number; let tails represent an odd number. Let each toss represent one trial. Toss 52 times. The number of heads generated divided by 52 represents the probability that she starts with sit-ups.

3. Possible answer: Use a coin as a random generator. Let heads represent red; let tails represent black. Each toss of the coin would represent one trial. Generate 20 trials and record the results.

4. Possible answer: Use a number cube as a random generator. Let a result of 1 represent red, 2 represent green, 3 represent yellow, 4 represent blue, 5 represent tan, and 6 represent dark brown. Each roll of the cube represents one trial. Generate 30 trials and record the results.

5. Possible answer: Use a number cube and a coin as random generators. A result of 1 or 2 could represent rain on Saturday and a result of 3, 4, or 5 could mean no rain on Saturday. A result of 6 would mean roll again. A coin could be flipped for Sunday's results, using heads for rain and tails for no rain. Each pair of results would represent one trial. Generate 20 trials and record the results.

Lesson 4.3

1. Mean: $48.\overline{3}$
   Median: 48
   Mode: 35

2. Mean: $79.5\overline{3}$
   Median: 89
   Mode: 95

3.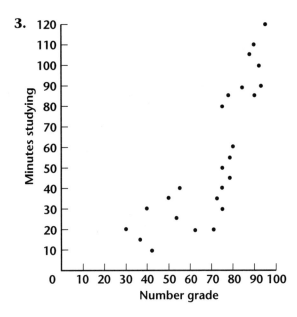

4.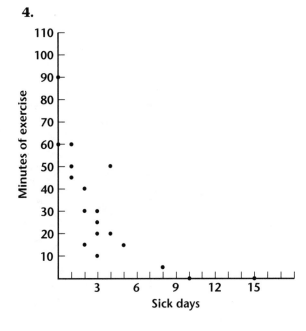

ANSWERS

Lesson 4.4

1.

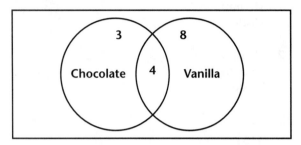

2.

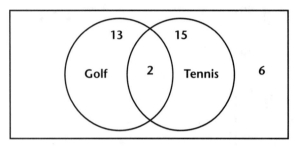

3. a. 135   **b.** 126   **c.** 91   **d.** 170

Lesson 4.5

1. 60

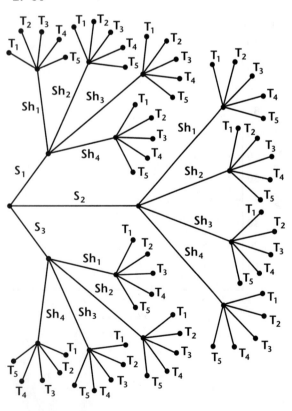

2. 36

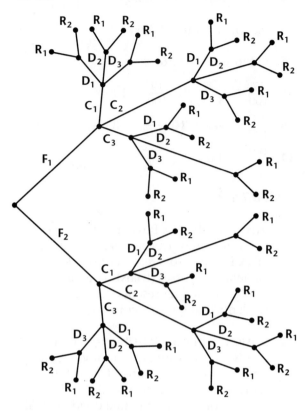

3. 6760   **4.** 100   **5.** 12

Lesson 4.6

1. $\frac{3}{12} = \frac{1}{4} = 0.25 = 25\%$

2. $\frac{7}{11} = 0.\overline{63} \approx 63.63\%$

3. $\frac{16}{99} = 0.\overline{16} \approx 16.16\%$

4. $\frac{78}{145}$ **5.** $\frac{49}{145}$ **6.** $\frac{73}{145}$ **7.** $\frac{102}{145}$

Lesson 4.7

1. dependent   **2.** independent

3. dependent   **4.** independent

5. $\frac{1}{4} = 25\%$ **6.** $\frac{1}{16} = 6.25\%$

7. $\frac{1}{12} \approx 8.3\%$ **8.** $\frac{1}{100} = 1\%$

ANSWERS

Reteaching—Chapter 5

Lesson 5.1

1. function   **2.** not a function   **3.** function

4. function   **5.** 38   **6.** −8   **7.** 57   **8.** $\frac{1}{9}$

9. not a function   **10.** function

Lesson 5.2

1. parent function: $y = |x|$
reflection through the x-axis

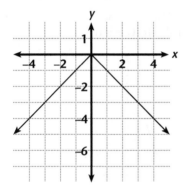

2. parent function: $y = |x|$
stretched by a factor of 3

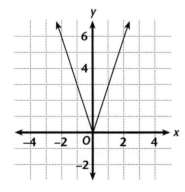

3. parent function: $y = |x|$
shifted 3 units to the right

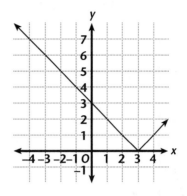

4. parent function: $y = |x|$
shifted down 1 unit

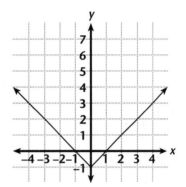

5. parent function: $y = |x|$
reflected through the x-axis, stretched by a
factor of 2, and shifted 1 unit to the left

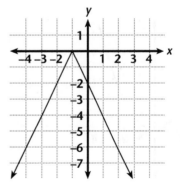

6. parent function: $y = |x|$
reflected through the x-axis, stretched by a
factor of 3 and shifted down 3 units

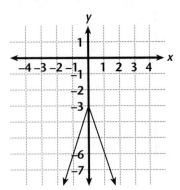

7. parent function: $y = |x|$
shifted down 3 units

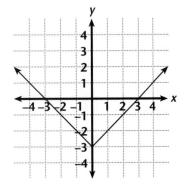

8. parent function: $y = x^2$
stretched by a factor of $\frac{1}{2}$

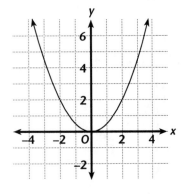

9. parent function: $y = |x|$
stretched by a factor of 2 and shifted 3
units to the right

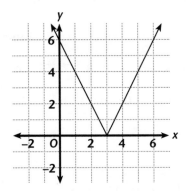

10. parent function: $y = x^2$
shifted 3 units to the left

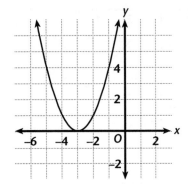

Lesson 5.3

1. $y = \frac{1}{x}$; scale factor of 3

2. $y = x^2$; scale factor of $\frac{1}{5}$

3. $y = |x|$; scale factor of 3

4. $y = 10^x$; scale factor of 2

ANSWERS

5. The graph is stretched vertically by 3; it rises more sharply.

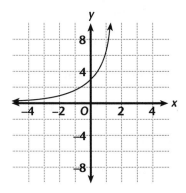

6. The graph is stretched vertically by $\frac{1}{4}$; it is more spread out.

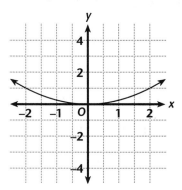

7. 2; stretched by a factor of 2

8. 3; stretched by a factor of 3

9. 5; stretched by a factor of 5

10. $\frac{1}{3}$; stretched by a factor of $\frac{1}{3}$

11. $\frac{1}{4}$; stretched by a factor of $\frac{1}{4}$

12. $\frac{1}{5}$; stretched by a factor of $\frac{1}{5}$

13. $\frac{2}{5}$; stretched by a factor of $\frac{2}{5}$

14. 0.65; stretched by a factor of 0.65

Lesson 5.4

1. yes   **2.** no   **3.** no   **4.** no   **5.** yes

6. yes   **7.** maximum; $(0, 0)$

8. minimum; $(0, -2)$ **9.** maximum; $(0, -2)$

10. minimum; $(0, 0)$ **11.** minimum; $(0, -3)$

12. maximum; $(0, -1)$

Lesson 5.5

1. $(-2, 4)$ **2.** $(9, 9)$ **3.** $(-10, 11)$

4. $(3, 2)$ **5.** $(13, 23)$ **6.** $(-17, 9)$

7. $y = 2^x$; vertical translation 3 units up

8. $y = \frac{1}{x}$; vertical translation 2 units down

9. $y = x^2$; vertical translation 2 units up

10. $y = |x|$; vertical translation 4 units down

11. $y = 10^x$; horizontal translation 2 units to the right

12. $y = |x|$; horizontal translation 3 units to the left

13. $y = x^2$; horizontal translation 4 units to the right

14. $y = x^2$; vertical translation 3 units down

Lesson 5.6

1. $y = |x|$ **2.** $y = x^2$ **3.** $y = 2^x$ **4.** $y = x$

5. $y = \frac{1}{x}$ **6.** $y = \frac{1}{x}$ **7.** $y = x^2$ **8.** $y = x^2$

9. The parent function has been shifted 2 units to the left, stretched by a factor of 3, reflected through the x-axis and shifted 3 units down.

10. The parent function has been shifted 3 units to the right, stretched by a factor of 2, and shifted 1 unit up.

11. The parent function has been shifted 1 unit to the right, stretched by a factor of 4, and shifted 2 units up.

12. The parent function has been stretched by a factor of 4 and shifted down 2 units.

ANSWERS

Reteaching—Chapter 6

Lesson 6.1

1. 1,000,000   2. 1,000,000,000   3. 625

4. 243   5. 64   6. 256   7. 3^4 8. 4^6

9. 6^7 10. 10^4 11. 5^3 12. 2^6

13. $4 \times 10^4 + 3 \times 10^3 + 7 \times 10^2 + 9 \times 10^1 + 5$

14. $1 \times 10^5 + 4 \times 10^4 + 5 \times 10^3 + 2 \times 10^2 + 8 \times 10^1 + 6$

15. $7 \times 10^5 + 2 \times 10^4 + 3 \times 10^3 + 6 \times 10^2 + 2 \times 10^1 + 1$

16. $3 \times 10^4 + 6 \times 10^3 + 2 \times 10^2 + 9 \times 10^1 + 1$

17. 4^8 18. 3^{11} 19. 3^8 20. 7^3 21. 4^8

22. 5^6 23. 3^{15} 24. 9^{12} 25. 2^{15} 26. 10^2

27. 7^{21} 28. 5^4 29. 3^{50} 30. 6^{10}

Lesson 6.2

1. no   2. yes   3. yes   4. yes   5. no

6. no   7. no   8. yes   9. yes   10. yes

11. yes   12. yes   13. no   14. no   15. no

16. yes   17. no   18. yes   19. no   20. yes

21. $80x^{12}y^8$ 22. $-1024m^{15}n^5$ 23. a^8b^9

24. $-4000a^{11}b^{10}$ 25. $-108m^5n^8$

26. $288{,}000\, a^{15}b^{16}$ 27. $-12r^7s^{11}t^8$

28. $-32x^{10}y^7z^4$ 29. $2048r^{10}s^{11}t^8$

30. $-13{,}500\, m^{18}n^{10}$ 31. $-48x^{13}y^9$

32. $-288a^{24}b^{11}$ 33. $-512r^{10}s^6t^7$

34. $16m^{15}n^{13}$ 35. $-16a^{23}b^{19}$

36. $-108x^{16}y^{13}$

Lesson 6.3

1. 1   2. $\frac{1}{25}$ 3. 1   4. $\frac{1}{4}$ 5. $\frac{1}{27}$ 6. 1

7. $\frac{1}{125}$ 8. $\frac{1}{64}$ 9. a^{-2} 10. c^{-5} 11. y^{-3}

12. m^{-9} 13. p^8 14. q^{-5} 15. x^{-11}

16. z^3 17. t^5 18. 5^5 or 3125   19. x^{-5}

20. $\frac{1}{3}$ 21. t^{-3} 22. 1,048,576   23. 25

24. a^{-3} 25. r^3 26. $\frac{1}{1024}$

Lesson 6.4

1. 4.57×10^9 2. 2.3×10^{-6}

3. 4.58×10^{-3} 4. 6.2×10^7

5. 7.05×10^{10} 6. 8.75×10^{-5}

7. 5.8×10^3 8. 2.6×10^{-2} 9. 3.5×10^7

10. 7.2×10^{-8} 11. 2.07×10^{12}

12. 3.05×10^{-3} 13. 2.4×10^8

14. 5×10^2 15. 2.236×10^7 16. 4×10^4

17. 2.3646 E 08   18. 3 E 02   19. 2.686 E 12

20. 2 E 02

Lesson 6.5

1. The function increases as x increases.

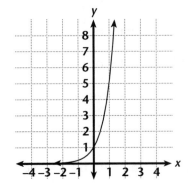

ANSWERS

2. The function decreases as x increases.

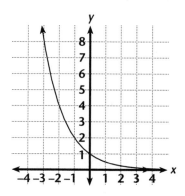

3. The function increases as x increases.

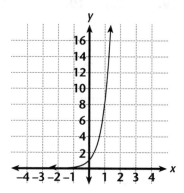

4. The function decreases as x increases.

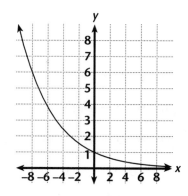

5. $y = 3^x$ **6.** $y = \left(\frac{1}{3}\right)^x$ **7.** $y = 4^x$

8. $y = \left(\frac{1}{4}\right)^x$

Lesson 6.6

1. $V = P(1 - r)^t$; V = present value,
P = original value, r = rate of change,
t = number of years
$V = 24(1 - .06)^5$

2. $P = N(1 + r)^t$; P = present population,
N = original population,
r = rate of change, t = number of years
$500{,}000 = N(1 + .02)^6$ or
$P = 500{,}000(1 + .02)^{-6}$

3. $V = P(1 + r)^t$; V = present value,
P = original value, r = rate of change,
t = number of years
$V = 54(1 + .04)^7$

4. $9030.56   **5.** about 29,364   **6.** 6480

Reteaching—Chapter 7

Lesson 7.1

1. 17   **2.** 39   **3.** -47 **4.** -42 **5.** 4

6. 1   **7.** 5   **8.** 72   **9.** 4   **10.** 3   **11.** -12

12. -72 **13.** $\frac{1}{3}$ **14.** $5\frac{1}{6}$ **15.** 3   **16.** $-\frac{17}{36}$

17. 18   **18.** -24 **19.** 46   **20.** 5

21. 75 cubic meters   **22.** 110 square meters

23. 125 cubic feet   **24.** 475 square feet

Lesson 7.2

1. $6b^2 + b$ **2.** $10c^2 + c$ **3.** $5b^3 - 3b^2 + 7b$

4. $5y^3 + 5y^2 + 6y - 1$

5. $2r^3 + 6r^2 + 7r + 6$

6. $7m^3 - 5m^2 - 4m - 5$ **7.** $-3x^2 + 4x - 3$

8. $4x^2 - 2x + 9$ **9.** $3x - 1$ **10.** $7x^3 + 1$

11. $-3c^2 - c - 5$ **12.** $-n^2 + 2n - 3$

13. $2z^2 + z + 1$ **14.** $-5r^2 - 4r + 9$

15. $-4t^2 + t$ **16.** $9q^2 + q + 3$

17. $-5 + 2a + 3a^2$ **18.** $-5e^3 + 4e^2 - 2e$

19. $-2x^2 + 2x + 8$ **20.** $-8x + 3$

21. $6x^2 - x + 3$ **22.** $x^2 + 3x - 7$

ANSWERS

23. $3x^2 + 2$ **24.** $-x + 6$

Lesson 7.3

1. $2x + 6$ **2.** $6x - 6$ **3.** $x^2 - 3x$

4. $-x^2 + x$ **5.** $2x^2 + 4x$ **6.** $3x^2 - 3x$

7. $2x^2 - 4x$ **8.** $6x^2 + 4x$ **9.** $x^2 + 3x + 2$

10. $x^2 + x - 2$ **11.** $x^2 - x - 2$

12. $x^2 - 3x + 2$ **13.** $x^2 + 6x + 9$

14. $x^2 - 6x + 9$ **15.** $x^2 - 9$ **16.** $x^2 - 9$

17. $2x^2 + 5x + 3$ **18.** $2x^2 - 5x + 2$

19. $3x^2 - 4x - 4$ **20.** $6x^2 - x - 2$

Lesson 7.4

1. $4x + 20$ **2.** $5x - 10$ **3.** $2x^2 - 2x$

4. $6x^2 + 2x$ **5.** $-5x^2 + 30x$ **6.** $3x^2 + 9x$

7. $x^2 + 5x + 4$ **8.** $x^2 + 5x + 6$

9. $x^2 + 2x - 15$ **10.** $2x^2 + 7x + 6$

11. $3x^2 - 18x + 15$ **12.** $12x^2 - 25x + 12$

13. $x^2 + 7x + 10$ **14.** $x^2 - x - 12$

15. $x^2 - 8x + 15$ **16.** $2x^2 + 3x - 9$

17. $4x^2 - 20x + 25$ **18.** $25x^2 - 40x + 16$

19. $4x^2 - 3x - 1$ **20.** $2x^2 - 7x + 6$

21. $-6x^2 + 17x - 12$

Lesson 7.5

1. 1, 2, 3, 4, 6, 12; no   **2.** 1, 5, 7, 35; no

3. 1, 47; yes   **4.** 1, 3, 19, 57; no

5. 1, 7, 11, 77; no   **6.** 1, 97; yes

7. $3m(m - 7)$ **8.** $t(8t + 15)$

9. $3(6p^2 + 7p + 3)$ **10.** $4d(d^2 - 5d + 2)$

11. $(x + 8)(x + 5)$ **12.** $(x - 7)(x - 3)$

13. $(x + 3)(x + 4)$ **14.** $(x - 2)(x - 3)$

15. $(x + 5)(x - 2)$ **16.** $(x - 4)(x + 1)$

Lesson 7.6

1. $x^2 + 8x + 16$ **2.** $4v^2 - 20v + 25$

3. $9d^2 - 6ad + a^2$ **4.** $100k^2 - 20kt + t^2$

5. $(x + 1)^2$ **6.** prime   **7.** $(2p + 4)^2$

8. $(2a - b)^2$ **9.** $(r - 2s)^2$ **10.** $(7b - 3c)^2$

11. $t^2 - 25$ **12.** $m^2 - d^2$ **13.** $9r^2 - s^4$

14. $25p^2 - 16q^2$ **15.** $(t + 7)(t - 7)$

16. $(2 - a)(2 + a)$ **17.** prime

18. $(2s - t)(2s + t)$ **19.** $(5c - 2q)(5c + 2q)$

20. $(4b + c^2)(4b - c^2)$ **21.** $(mn + p)(mn - p)$

22. $(s^2 + t^2)(s + t)(s - t)$

Lesson 7.7

1. $(x + 1)(x + 2)$ **2.** $(x - 3)(x + 4)$

3. $(x - 7)(x - 3)$ **4.** $(x + 5)(x - 1)$

5. $(x + 8)(x + 3)$ **6.** $(x - 4)(x - 4)$

7. $(x - 2)(x + 1)$ **8.** $(x + 4)(x - 1)$

9. $(x + 3)(x + 1)$ **10.** $(x - 3)(x - 1)$

11. $(x + 4)(x - 2)$ **12.** $(x + 5)(x - 4)$

13. $(x + 5)(x - 3)$ **14.** prime

15. $(x + 3)(x - 4)$ **16.** $(x + 4)(x + 2)$

17. $(x - 2)(x - 18)$ **18.** $(x + 6)(x - 4)$

ANSWERS

Reteaching—Chapter 8

Lesson 8.1

1. vertex: (5, 1); axis of symmetry: $x = 5$; minimum: 1

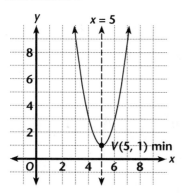

2. vertex: (4, −2); axis of symmetry: $x = 4$; maximum: −2

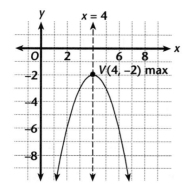

3. vertex: (−1, 4); axis of symmetry: $x = -1$; maximum: 4

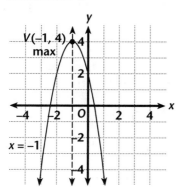

4. vertex: (3, 5); axis of symmetry: $x = 3$; minimum: 5

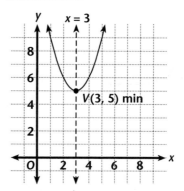

5. 1 and 2

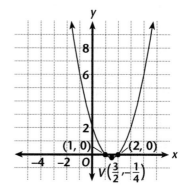

6. −2 and 5

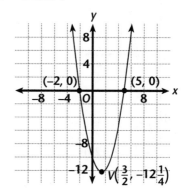

7. −3 and −1

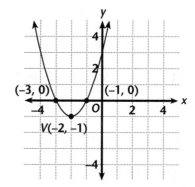

ANSWERS

8.

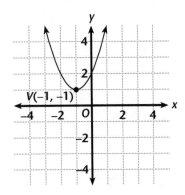

9. $\frac{1}{2}$ and 2

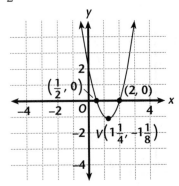

10. -2 and $\frac{3}{2}$

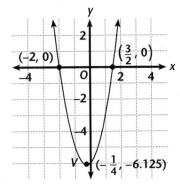

Lesson 8.2

1. 6   **2.** -14 **3.** -16.58 **4.** 22

5. 15.49   **6.** -29 **7.** ± 5 **8.** ± 8.66

9. ± 10.39 **10.** -7 and 5   **11.** -6 and 14

12. -11 and 5   **13.** -16 and 2

14. approximately -3.92 and 7.92

15. approximately -10.18 and 12.18

Lesson 8.3

1. $36; -36$ **2.** $\frac{25}{4}; -\frac{25}{4}$ **3.** $100; -100$

4. $1; -1$ **5.** $25; -25$ **6.** $\frac{121}{4}; -\frac{121}{4}$

7. $y = (x + 1)^2 + 0; V(-1, 0)$

8. $y = (x - 4)^2 - 13; V(4, -13)$

9. $y = (x + 2)^2 - 7; V(-2, -7)$

10. $y = (x - 1)^2 + 3; V(1, 3)$

11. $y = (x - 6)^2 - 72; V(6, -72)$

12. $y = (x + 1)^2 - 5; V(-1, -5)$

13. $y = \left(x - \frac{3}{2}\right)^2 + \frac{15}{4}; V\left(\frac{3}{2}, \frac{15}{4}\right)$

14. $y = \left(x - \frac{5}{2}\right)^2 - \frac{125}{4}; V\left(\frac{5}{2}, -\frac{125}{4}\right)$

15. $y = \left(x + \frac{1}{2}\right)^2 + \frac{3}{4}; V\left(-\frac{1}{2}, \frac{3}{4}\right)$

Lesson 8.4

1. -2 and 1   **2.** -5 and 2   **3.** -2 and -3

4. -1 **5.** $5 + \sqrt{22}$ and $5 - \sqrt{22}$

6. $-2 + \sqrt{7}$ and $-2 - \sqrt{7}$ **7.** -4 and 5

8. 3 and 4   **9.** -5 and 1   **10.** -2 and -4

11. 0 and -6 **12.** $-3 + \sqrt{11}$ and $-3 - \sqrt{11}$

Lesson 8.5

1. 1 and 4   **2.** -4 and 6   **3.** -3

4. -5 and 2   **5.** $-\frac{3}{2}$ and 2

6. $\frac{-1 \pm \sqrt{33}}{4}$, or approximately -1.69
and 1.19

7. -4 and 2   **8.** $-\frac{5}{2}$ and 3   **9.** $\frac{1}{2}$ and $\frac{3}{2}$

10. 0; one solution   **11.** 9; two solutions

12. -7; no real solutions

ANSWERS

Lesson 8.6

1. $-2 < x < -1$

2. $x \le 3$ or $x \ge 5$

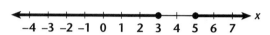

3. $x < -3$ or $x > 3$

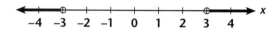

4. $-3 \le x \le 2$

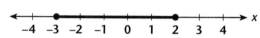

5.

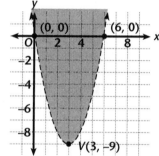

6.

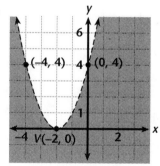

7.

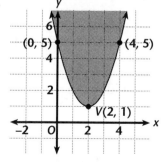

8.

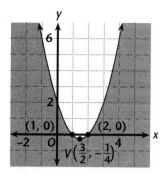

Reteaching—Chapter 9

Lesson 9.1

1. 4.24   **2.** 10.72   **3.** -3.10 **4.** 11.58

5. -0.28 **6.** 38.73   **7.** 37 feet

8. 15.6 meters   **9.** about 316.23 feet

10. $y = \sqrt{x + 4}$; the graph of the parent function is shifted horizontally 4 units to the left.

11. $y = 2\sqrt{x} - 1$; the graph of the parent function is shifted vertically 1 unit down and stretched by a factor of 2.

12. $y = -\sqrt{x + 2}$; the graph of the parent function is shifted horizontally 2 units to the left and reflected through the x-axis.

13. $y = \sqrt{x - 3} + 2$; the graph of the parent function is shifted horizontally 3 units to the right and vertically 2 units up.

Lesson 9.2

1. $4\sqrt{3}$ **2.** $10b\sqrt{2}$ **3.** $\frac{\sqrt{2c}}{c}, c \ne 0$

4. $\frac{a\sqrt{15}}{5}$ **5.** $7\sqrt{6}$ **6.** $6\sqrt{3}$ **7.** $12\sqrt{2}$

8. $6\sqrt{2}$ **9.** $-4\sqrt{5}$ **10.** $4\sqrt{2}$ **11.** $31\sqrt{3}$

12. $-\sqrt{6} - 6$ **13.** $2\sqrt{3} + 5\sqrt{2}$

14. $5\sqrt{3} - 6$ **15.** $3\sqrt{2} + 2\sqrt{3}$ **16.** -1

17. $12 - 4\sqrt{2} + 6\sqrt{3} - 2\sqrt{6}$ **18.** $8 + 2\sqrt{15}$

ANSWERS

Lesson 9.3

1. 10 **2.** 11 **3.** 23 **4.** 8 **5.** 5 **6.** 2

7. $\pm 4\sqrt{15}$ **8.** 25 **9.** 2 **10.** 2

11. $3, -5$ **12.** $-3 \pm 2\sqrt{5}$ **13.** 12

14. $2, 0$ **15.** 1

Lesson 9.4

1. 5 **2.** 12 **3.** 15 **4.** about 8.49 inches

5. about 8.7 meters **6.** 10 miles

7. 75 miles **8.** 27.73 meters **9.** 10.61 feet

Lesson 9.5

1. 10 **2.** 9 **3.** $\sqrt{65}$, or ≈ 8.06

4. The length of LM is $\sqrt{50}$. The length of MN is $\sqrt{18}$. The length of LN is $\sqrt{68}$. Since $(\sqrt{50})^2 = 50$, $(\sqrt{18})^2 = 18$, $(\sqrt{68})^2 = 68$, and $50 + 18 = 68$, triangle LMN is a right triangle.

5. $M\left(-\frac{7}{2}, 2\right)$ **6.** $M\left(-3, -\frac{5}{2}\right)$ **7.** $M(-2, -4)$

8. $E(10, 4)$ **9.** $C(1, -1)$

Lesson 9.6

1. $x^2 + y^2 = 25$ **2.** $x^2 + y^2 = 6.25$

3. $x^2 + y^2 = \frac{9}{16}$ **4.** $C(-5, -1); r = 6$

5. $C(4, 1); r = 3\sqrt{3}$ **6.** $C(3, 3); r = 2.8$

7. $BC = 3\sqrt{5}$, $CA = 3\sqrt{5}$

8. $AC = 2\sqrt{13}$, $BD = 2\sqrt{13}$

9. $AC = \sqrt{(0 - a)^2 + (0 - a)^2} = \sqrt{a^2 + a^2} = \sqrt{2a^2} = a\sqrt{2}$

$BD = \sqrt{(a - 0)^2 + (0 - a)^2} = \sqrt{a^2 + a^2} = \sqrt{2a^2} = a\sqrt{2}$

Lesson 9.7

1. 1.333 **2.** 0.577 **3.** 1.875 **4.** 0.364

5. 1 **6.** 3.078 **7.** 9.4 feet **8.** 23.3 meters

9. 10.4 feet

Lesson 9.8

1. $\sin A = 0.6$; $\cos A = 0.8$

2. $\sin A = 0.471$; $\cos A = 0.882$

3. $\sin A = 0.667$; $\cos A = 0.745$

4. $\sin 25° = 0.423$; $\cos 25° = 0.906$

5. $\sin 45° = 0.707$; $\cos 45° = 0.707$

6. $\sin 90° = 1$; $\cos 90° = 0$

7. 28.7 feet **8.** 388.2 meters **9.** 319.3 feet

Reteaching—Chapter 10

Lesson 10.1

1. -1 **2.** -6 and 1 **3.** -3 and 3

4. $-\frac{17}{3}$ and $\frac{1}{5}$ **5.** $\frac{1}{6}$ and $\frac{1}{2}$

6. $\frac{1}{6}$ and undefined

7. shift down 3 units

8. shift to the right 1 unit

9. shift up 2 units and to the left 1 unit

10. stretch by a factor of 2 and shift to the left 2 units

11. stretch by a factor of 3, shift to the left 5 units and up 2 units

12. shift down 5 units and to the right 3 units

Lesson 10.2

1. yes **2.** yes **3.** no **4.** 8 **5.** 10 **6.** 5

ANSWERS

7. 24 inches **8.** 45 amps **9.** $10\frac{2}{3}$ days

10. 25 meters

Lesson 10.3

1. $\frac{2t}{t-1}$, $t \neq 1$ **2.** $\frac{m+3}{2(m-2)}$, $m \neq 2$

3. $\frac{m}{2+3m}$, $m \neq 0$ and $m \neq -\frac{2}{3}$

4. $\frac{3}{y+2}$, $y \neq -2$ **5.** $\frac{3}{x+1}$, $x \neq -1$

6. $\frac{4}{r-3}$, $r \neq 3$ **7.** $m+2$, $m \neq -1$

8. $\frac{1}{a-3}$, $a \neq 2$ and $a \neq 3$ **9.** $\frac{3}{b+2}$, $b \neq -2$

10. $\frac{r}{r+1}$, $r \neq 1$ and $r \neq -1$

11. $\frac{x+1}{x}$, $x \neq 0$ and $x \neq 1$

12. $\frac{x+3}{x-4}$, $x \neq 2$ and $x \neq 4$

13. $x-2$

Lesson 10.4

1. $2t$, $t \neq 0$ and $t \neq 1$ **2.** $\frac{2}{3}$, $a \neq 0$ and $a \neq \frac{3}{2}$

3. $\frac{c+3}{2}$, $c \neq -3$

4. $\frac{4}{d+5}$, $d \neq 0$, $d \neq -3$ and $d \neq -5$

5. $2(x-3)$, $x \neq -3$

6. $y^2 - 1$, $y \neq -2$ and $y \neq 2$ **7.** $\frac{13x}{20}$

8. $\frac{37}{6b}$, $b \neq 0$ **9.** $-\frac{7}{12t}$, $t \neq 0$

10. $\frac{5m}{2(m-1)}$, $m \neq 1$ **11.** $\frac{c}{2(c-4)}$, $c \neq 4$

12. $\frac{3y^2+2y}{(y+2)(y-2)}$, $y \neq -2$ and $y \neq 2$

Lesson 10.5

1. -6 **2.** $-\frac{6}{13}$ and 2, $d \neq 0$

3. 3, $h \neq 0$ and $h \neq 6$

4. -3, $x \neq 4$ and $x \neq 2$

5. $2\frac{2}{5}$ hours, or 2 hours and 24 minutes

6. $7\frac{1}{2}$ hours, or 7 hours and 30 minutes

Lesson 10.6

1. 3 **2.** -10 **3.** -3 **4.** 0 and 6

5. -1 and 6 **6.** $\frac{4}{5}$ **7.** 300 miles

8. 80 degrees **9.** 6 friends

10. 22.5 degrees, 67.5 degrees

Lesson 10.7

1. $3x - 6 = 15$ Given
$3x = 21$ Addition Property
$x = 7$ Division Property

2. $8x^2 = 72$ Given
$x^2 = 9$ Division Property
$x = \pm 3$ Property of Square Roots

3. $x^2 - x = 12$ Given
$x^2 - x - 12 = 0$ Subtraction Property
$(x-4)(x+3) = 0$ Factoring
$x = 4$ or $x = -3$ Addtion and Subtraction Property

4. $\frac{5}{x} = \frac{x-3}{2}$ Given
$2(5) = x(x-3)$ Cross products
$10 = x^2 - 3x$ Distribution Property
$0 = x^2 - 3x - 10$ Subtraction Property
$0 = (x-5)(x+2)$ Factoring
$x = 5$ or $x = -2$ Addition and Subtraction Property

5. Let x and y represent any integers, where $x \neq y$.   Given
$2x$ and $2y$ are even numbers.   Any integer in the form $2x$ is even.
$2x + 1$ and $2y + 1$ are odd.   Any odd integer is an even number plus 1.
$(2x + 1) + (2y + 1) =$ Add any two odd integers.
$= 2x + 2y + 1 + 1$ Commutative Property

$= 2x + 2y + 2$ Add like terms.

$= 2(x + y + 1)$ Distributive Property

The integer represented by $2(x + y + 1)$ is an even number because any integer times 2 is even. Thus, any two odd integers added together will always result in an even integer.

6. Let x represent any integer.   Given

$2x$ is an even number.   Any integer in the form $2x$ is even.

$2x + 1$ is odd.   Any odd integer is an even number plus 1.

$(2x)(2x + 1) =$ Multiplication

$= 4x^2 + 2x$ Distributive Property

$= 2(x^2 + x)$ Distributive Property

The integer represented by $2(x^2 + x)$ is an even number because any integer times 2 is even. Thus, the product of any odd number and any even number will always result in an even number.

7. Let x represent any integer.   Given

$2x$ is an even integer.   Any integer in the form $2x$ is even.

$2x + 1$ is odd.   Any odd integer is an even number plus 1.

$2x + 2x + 1 =$ Addition

$= 4x + 1$ Add like terms.

$= 2(2x) + 1$ Distributive Property

The integer represented by $2(2x) + 1$ is odd because it is an even number, $2(2x)$, plus 1. Thus, any odd integer plus any even integer will always result in an odd integer.

8. Let x represent any integer.   Given

$2x$ is an even number.   Any integer in the form $2x$ is even.

$2x + 1$ is odd.   Any odd integer is an even number plus 1.

$(2x + 2x + 1)^2 =$ Squaring of the sum of an even number and an odd number

$= (4x + 1)^2$ Add like terms.

$= 16x^2 + 8x + 1$ Square a binomial.

$= 2(8x^2 + 4x) + 1$ Distributive Property

The integer represented by $2(8x^2 + 4x) + 1$ is odd because it is an even number, $2(8x^2 + 4x)$, plus 1. Thus, the square of the sum of any two consecutive integers will always be an odd number.

9. Let x represent any integer.   Given

Let y represent any integer.   Given

$3x$ and $3y$ are each multiples of 3.   Any integer times 3 is a multiple of 3.

$3x + 3y =$ Addition

$= 3(x + y)$ Distributive Property

The integer represented by $3(x + y)$ is a multiple of 3 because any number times 3 is a multiple of 3. Thus, any two multiples of 3 added together will always result in a multiple of 3.

10. Let x represent any integer.   Given

Let y represent any integer.   Given

$2x$ is an even number.   Any integer in the form $2x$ is even.

$5y$ is a multiple of 5.   Any integer in the form $5y$ is a multiple of 5.

$(2x)(5y) =$ Multiplication

$= 10xy$ Multiplication Property

$= 10(xy)$ Distributive Property

The integer represented by $10(xy)$ is divisible by 10 because any number times 10 is divisible by 10. Thus, the product of any even integer and any multiple of 5 is divisible by 10.